THE CHEMISTRY OF EXPLOSIVES

Second Edition

RSC Paperbacks

RSC Paperbacks are a series of inexpensive texts suitable for teachers and students and give a clear, readable introduction to selected topics in chemistry. They should also appeal to the general chemist. For further information on all available titles contact:

Sales and Customer Care Department, Royal Society of Chemistry, Thomas Graham House, Science Park, Cambridge CB4 4WF, UK
Telephone: + 44 (0)1223 432360
Fax: + 44 (0)1223 426017
E-mail: sales@rsc.org

Recent Titles Available

The Science of Chocolate
By Stephen T. Beckett
The Science of Sugar Confectionery
By W.P. Edwards
Colour Chemistry
By R.M. Christie
Beer: Quality, Safety and Nutritional Aspects
By P.S. Hughes and E.D. Baxter
Understanding Batteries
By Ronald M. Dell and David A.J. Rand
Principles of Thermal Analysis and Calorimetry
Edited by P.J. Haines
Food: The Chemistry of its Components (Fourth Edition)
By Tom P. Coultate
Green Chemistry: An Introductory Test
By Mike Lancaster
The Misuse of Drug Acts: A Guide for Forensic Scientists
By L.A. King
Chemical Formulation: An Overview of Surfactant-based Chemical Preparations in Everyday Life
By A.E. Hargreaves
Life, Death and Nitric Oxide
By Antony Butler and Rosslyn Nicholson
A History of Beer and Brewing
By Ian S. Hornsey

Future titles may be obtained immediately on publication by placing a standing order for RSC Paperbacks. Information on this is available from the address above.

RSC Paperbacks

THE CHEMISTRY OF EXPLOSIVES

Second Edition

JACQUELINE AKHAVAN

Department of Environmental and Ordnance Systems
Cranfield University
Royal Military College of Science
Swindon SN6 8LA

advancing the chemical sciences

WARNING STATEMENT

It is both dangerous and illegal to participate in unauthorized experimentation with explosives.

ISBN 0-85404-640-2

A catalogue record for this book is available from the British Library

Published by The Royal Society of Chemistry, Thomas Graham House, Science Park, Milton Road, Cambridge CB4 0WF, UK
For further information visit our web site at www.rsc.org

Typeset by Vision Typesetting Ltd, Manchester
Printed in Great Britain by Henry Ling Limited, at the Dorset Press, Dorchester, DT1 1HD.

Preface

This book outlines the basic principles needed to understand the mechanism of explosions by chemical explosives. The history, theory and chemical types of explosives are introduced, providing the reader with information on the physical parameters of primary and secondary explosives. Thermodynamics, enthalpy, free energy and gas equations are covered together with examples of calculations, leading to the power and temperature of explosions. A very brief introduction to propellants and pyrotechnics is given, more information on these types of explosives should be found from other sources. This second edition introduces the subject of Insensitive Munitions (IM) and the concept of explosive waste recovery. Developments in explosive crystals and formulations have also been updated. This book is aimed primarily at 'A' level students and new graduates who have not previously studied explosive materials, but it should prove useful to others as well. I hope that the more experienced chemist in the explosives industry looking for concise information on the subject will also find this book useful.

In preparing this book I have tried to write in an easy to understand style guiding the reader through the chemistry of explosives in a simple but detailed manner. Although the reader may think this is a new subject he or she will soon find that basic chemistry theories are simply applied in understanding the chemistry of explosives.

No book can be written without the help of other people and I am aware of the help I have received from other sources. These include authors of books and journals whom I have drawn upon in preparing this book. I am also grateful for the comments from the reviewers of the first edition of this book.

I would particularly like to thank my husband Shahriar, who has always supported me.

Contents

Chapter 3
Combustion, Deflagration and Detonation   49

Chapter 4
Ignition, Initiation and Thermal Decomposition 63

Chapter 5
Thermochemistry of Explosives 74

Chapter 8
Introduction to Propellants and Pyrotechnics 149

Chapter 1

Introduction to Explosives

DEVELOPMENT OF BLACKPOWDER

Blackpowder, also known as gunpowder, was most likely the first explosive composition. In 220 BC an accident was reported involving blackpowder when some Chinese alchemists accidentally made blackpowder while separating gold from silver during a low-temperature reaction. According to Dr Heizo Mambo the alchemists added potassium nitrate [also known as saltpetre (KNO_3)] and sulfur to the gold ore in the alchemists' furnace but forgot to add charcoal in the first step of the reaction. Trying to rectify their error they added charcoal in the last step. Unknown to them they had just made blackpowder which resulted in a tremendous explosion.

Blackpowder was not introduced into Europe until the 13th century when an English monk called Roger Bacon in 1249 experimented with potassium nitrate and produced blackpowder, and in 1320 a German monk called Berthold Schwartz (although many dispute his existence) studied the writings of Bacon and began to make blackpowder and study its properties. The results of Schwartz's research probably speeded up the adoption of blackpowder in central Europe. By the end of the 13th century many countries were using blackpowder as a military aid to breach the walls of castles and cities.

Blackpowder contains a fuel and an oxidizer. The fuel is a powdered mixture of charcoal and sulfur which is mixed with potassium nitrate (oxidizer). The mixing process was improved tremendously in 1425 when the Corning, or granulating, process was developed. Heavy wheels were used to grind and press the fuels and oxidizer into a solid mass, which was subsequently broken down into smaller grains. These grains contained an intimate mixture of the fuels and oxidizer, resulting in a blackpowder which was physically and ballistically superior. Corned blackpowder gradually came into use for small guns and hand

1

grenades during the 15th century and for big guns in the 16th century.

Blackpowder mills (using the Corning process) were erected at Rotherhithe and Waltham Abbey in England between 1554 and 1603.

The first recording of blackpowder being used in civil engineering was during 1548–1572 for the dredging of the River Niemen in Northern Europe, and in 1627 blackpowder was used as a blasting aid for recovering ore in Hungary. Soon, blackpowder was being used for blasting in Germany, Sweden and other countries. In England, the first use of blackpowder for blasting was in the Cornish copper mines in 1670. Bofors Industries of Sweden was established in 1646 and became the main manufacturer of commercial blackpowder in Europe.

DEVELOPMENT OF NITROGLYCERINE

By the middle of the 19th century the limitations of blackpowder as a blasting explosive were becoming apparent. Difficult mining and tunnelling operations required a 'better' explosive. In 1846 the Italian, Professor Ascanio Sobrero discovered liquid nitroglycerine $[C_3H_5O_3(NO_2)_3]$. He soon became aware of the explosive nature of nitroglycerine and discontinued his investigations. A few years later the Swedish inventor, Immanuel Nobel developed a process for manufacturing nitroglycerine, and in 1863 he erected a small manufacturing plant in Helenborg near Stockholm with his son, Alfred. Their initial manufacturing method was to mix glycerol with a cooled mixture of nitric and sulfuric acids in stone jugs. The mixture was stirred by hand and kept cool by iced water; after the reaction had gone to completion the mixture was poured into excess cold water. The second manufacturing process was to pour glycerol and cooled mixed acids into a conical lead vessel which had perforations in the constriction. The product nitroglycerine flowed through the restrictions into a cold water bath. Both methods involved the washing of nitroglycerine with warm water and a warm alkaline solution to remove the acids. Nobel began to license the construction of nitroglycerine plants which were generally built very close to the site of intended use, as transportation of liquid nitroglycerine tended to generate loss of life and property.

The Nobel family suffered many set backs in marketing nitroglycerine because it was prone to accidental initiation, and its initiation in bore holes by blackpowder was unreliable. There were many accidental explosions, one of which destroyed the Nobel factory in 1864 and killed Alfred's brother, Emil. Alfred Nobel in 1864 invented the metal 'blasting cap' detonator which greatly improved the initiation of blackpowder. The detonator contained mercury fulminate $[Hg(CNO)_2]$ and was able to replace blackpowder for the initiation of nitroglycerine in bore holes. The mercury fulminate blasting cap produced an initial shock which was transferred to a separate container of nitroglycerine via a fuse, initiating the nitroglycerine.

After another major explosion in 1866 which completely demolished the nitroglycerine factory, Alfred turned his attentions into the safety problems of transporting nitroglycerine. To reduce the sensitivity of nitroglycerine Alfred mixed it with an absorbent clay, 'Kieselguhr'. This mixture became known as ghur dynamite and was patented in 1867.

Nitroglycerine (1.1) has a great advantage over blackpowder since it contains both fuel and oxidizer elements in the same molecule. This gives the most intimate contact for both components.

$$
\begin{array}{c}
\text{H} \\
\text{H}-\text{C}-\text{O}-\text{NO}_2 \\
\text{H}-\text{C}-\text{O}-\text{NO}_2 \\
\text{H}-\text{C}-\text{O}-\text{NO}_2 \\
\text{H}
\end{array}
$$

(1.1)

Development of Mercury Fulminate

Mercury fulminate was first prepared in the 17th century by the Swedish–German alchemist, Baron Johann Kunkel von Löwenstern. He obtained this dangerous explosive by treating mercury with nitric acid and alcohol. At that time, Kunkel and other alchemists could not find a use for the explosive and the compound became forgotten until Edward Howard of England rediscovered it between 1799 and 1800. Howard examined the properties of mercury fulminate and proposed its use as a percussion initiator for blackpowder and in 1807 a Scottish Clergyman, Alexander Forsyth patented the device.

DEVELOPMENT OF NITROCELLULOSE

At the same time as nitroglycerine was being prepared, the nitration of cellulose to produce nitrocellulose (also known as guncotton) was also being undertaken by different workers, notably Schönbein at Basel and Böttger at Frankfurt-am-Main during 1845–47. Earlier in 1833, Braconnot had nitrated starch, and in 1838, Pelouze, continuing the experiments of Braconnot, also nitrated paper, cotton and various other materials but did not realize that he had prepared nitrocellulose. With the announcement by Schönbein in 1846, and in the same year by

Böttger that nitrocellulose had been prepared, the names of these two men soon became associated with the discovery and utilization of nitrocellulose. However, the published literature at that time contains papers by several investigators on the nitration of cellulose before the process of Schönbein was known.

Many accidents occurred during the preparation of nitrocellulose, and manufacturing plants were destroyed in France, England and Austria. During these years, Sir Frederick Abel was working on the instability problem of nitrocellulose for the British Government at Woolwich and Waltham Abbey, and in 1865 he published his solution to this problem by converting nitrocellulose into a pulp. Abel showed through his process of pulping, boiling and washing that the stability of nitrocellulose could be greatly improved. Nitrocellulose was not used in military and commercial explosives until 1868 when Abel's assistant, E.A. Brown discovered that dry, compressed, highly-nitrated nitrocellulose could be detonated using a mercury fulminate detonator, and wet, compressed nitrocellulose could be exploded by a small quantity of dry nitrocellulose (the principle of a Booster). Thus, large blocks of wet nitrocellulose could be used with comparative safety.

DEVELOPMENT OF DYNAMITE

In 1875 Alfred Nobel discovered that on mixing nitrocellulose with nitroglycerine a gel was formed. This gel was developed to produce blasting gelatine, gelatine dynamite and later in 1888, ballistite, the first smokeless powder. Ballistite was a mixture of nitrocellulose, nitroglycerine, benzene and camphor. In 1889 a rival product of similar composition to ballistite was patented by the British Government in the names of Abel and Dewar called 'Cordite'. In its various forms Cordite remained the main propellant of the British Forces until the 1930s.

In 1867, the Swedish chemists Ohlsson and Norrbin found that the explosive properties of dynamites were enhanced by the addition of ammonium nitrate (NH_4NO_3). Alfred Nobel subsequently acquired the patent of Ohlsson and Norrbin for ammonium nitrate and used this in his explosive compositions.

Development of Ammonium Nitrate

Ammonium nitrate was first prepared in 1654 by Glauber but it was not until the beginning of the 19th century when it was considered for use in explosives by Grindel and Robin as a replacement for potassium nitrate in blackpowder. Its explosive properties were also reported in 1849 by

Reise and Millon when a mixture of powdered ammonium nitrate and charcoal exploded on heating.

Ammonium nitrate was not considered to be an explosive although small fires and explosions involving ammonium nitrate occurred throughout the world.

After the end of World War II, the USA Government began shipments to Europe of so-called Fertilizer Grade Ammonium Nitrate (FGAN), which consisted of grained ammonium nitrate coated with about 0.75% wax and conditioned with about 3.5% clay. Since this material was not considered to be an explosive, no special precautions were taken during its handling and shipment – workmen even smoked during the loading of the material.

Numerous shipments were made without trouble prior to 16 and 17 April 1947, when a terrible explosion occurred. The SS Grandchamp and the SS Highflyer, both moored in the harbour of Texas City and loaded with FGAN, blew up. As a consequence of these disasters, a series of investigations was started in the USA in an attempt to determine the possible causes of the explosions. At the same time a more thorough study of the explosive properties of ammonium nitrate and its mixtures with organic and inorganic materials was also conducted. The explosion at Texas City had barely taken place when a similar one aboard the SS Ocean Liberty shook the harbour of Brest in France on 28 July 1947.

The investigations showed that ammonium nitrate is much more dangerous than previously thought and more rigid regulations governing its storage, loading and transporting in the USA were promptly put into effect.

DEVELOPMENT OF COMMERCIAL EXPLOSIVES

Development of Permitted Explosives

Until 1870, blackpowder was the only explosive used in coal mining, and several disastrous explosions occurred. Many attempts were made to modify blackpowder; these included mixing blackpowder with 'cooling agents' such as ammonium sulfate, starch, paraffin, *etc.*, and placing a cylinder filled with water into the bore hole containing the blackpowder. None of these methods proved to be successful.

When nitrocellulose and nitroglycerine were invented, attempts were made to use these as ingredients for coal mining explosives instead of blackpowder but they were found not to be suitable for use in gaseous coal mines. It was not until the development of dynamite and blasting

gelatine by Nobel that nitroglycerine-based explosives began to domi-
nate the commercial blasting and mining industries. The growing use of
explosives in coal mining brought a corresponding increase in the
number of gas and dust explosions, with appalling casualty totals. Some
European governments were considering prohibiting the use of explosives in coal mines and resorting to the use of hydraulic devices or
compressed air. Before resorting to such drastic measures, some governments decided to appoint scientists, or commissions headed by them, to
investigate this problem. Between 1877 and 1880, commissions were
created in France, Great Britain, Belgium and Germany. As a result of
the work of the French Commission, maximum temperatures were set
for explosions in rock blasting and gaseous coal mines. In Germany and
England it was recognized that regulating the temperature of the explosion was only one of the factors in making an explosive safe and that
other factors should be considered. Consequently, a testing gallery was
constructed in 1880 at Gelsenkirchen in Germany in order to test the
newly-developed explosives. The testing gallery was intended to imitate
as closely as possible the conditions in the mines. A Committee was
appointed in England in 1888 and a trial testing gallery at Hebburn
Colliery was completed around 1890. After experimenting with various
explosives the use of several explosive materials was recommended,
mostly based on ammonium nitrate. Explosives which passed the tests
were called 'permitted explosives'. Dynamite and blackpowder both
failed the tests and were replaced by explosives based on ammonium
nitrate. The results obtained by this Committee led to the Coal Mines
Regulation Act of 1906. Following this Act, testing galleries were constructed at Woolwich Arsenal and Rotherham in England.

Development of ANFO and Slurry Explosives

By 1913, British coal production reached an all-time peak of 287 million
tons, consuming more than 5000 tons of explosives annually and by
1917, 92% of these explosives were based on ammonium nitrate. In
order to reduce the cost of explosive compositions the explosives industry added more of the cheaper compound ammonium nitrate to the
formulations, but this had an unfortunate side effect of reducing the
explosives' waterproofness. This was a significant problem because
mines and quarries were often wet and the holes drilled to take the
explosives regularly filled with water. Chemists overcame this problem
by coating the ammonium nitrate with various inorganic powders
before mixing it with dynamite, and by improving the packaging of the
explosives to prevent water ingress. Accidental explosions still occurred

involving mining explosives, and in 1950 manufacturers started to develop explosives which were waterproof and solely contained the less hazardous ammonium nitrate. The most notable composition was ANFO (Ammonium Nitrate Fuel Oil). In the 1970s, the USA companies Ireco and DuPont began adding paint-grade aluminium and monomethylamine nitrate (MAN) to their formulations to produce gelled explosives which could detonate more easily. More recent developments concern the production of emulsion explosives which contain droplets of a solution of ammonium nitrate in oil. These emulsions are waterproof because the continuous phase is a layer of oil, and they can readily detonate since the ammonium nitrate and oil are in close contact. Emulsion explosives are safer than dynamite, and are simple and cheap to manufacture.

DEVELOPMENT OF MILITARY EXPLOSIVES

Development of Picric Acid

Picric acid [(trinitrophenol) $(C_6H_3N_3O_7)$] was found to be a suitable replacement for blackpowder in 1885 by Turpin, and in 1888 blackpowder was replaced by picric acid in British munitions under the name Liddite. Picric acid is probably the earliest known nitrophenol: it is mentioned in the alchemical writings of Glauber as early as 1742. In the second half of the 19th century, picric acid was widely used as a fast dye for silk and wool. It was not until 1830 that the possibility of using picric acid as an explosive was explored by Welter.

Designolle and Brugère suggested that picrate salts could be used as a propellant, while in 1871, Abel proposed the use of ammonium picrate as an explosive. In 1873, Sprengel showed that picric acid could be detonated to an explosion and Turpin, utilizing these results, replaced blackpowder with picric acid for the filling of munition shells. In Russia, Panpushko prepared picric acid in 1894 and soon realized its potential as an explosive. Eventually, picric acid (1.2) was accepted all over the world as the basic explosive for military uses.

(1.2)

Picric acid did have its problems: in the presence of water it caused corrosion of the shells, its salts were quite sensitive and prone to accidental initiation, and picric acid required prolonged heating at high temperatures in order for it to melt.

Development of Tetryl

An explosive called tetryl was also being developed at the same time as picric acid. Tetryl was first prepared in 1877 by Mertens and its structure established by Romburgh in 1883. Tetryl (1.3) was used as an explosive in 1906, and in the early part of this century it was frequently used as the base charge of blasting caps.

H_3C NO_2

O_2N NO_2

NO_2

(1.3)

Development of TNT

Around 1902 the Germans and British had experimented with trinitrotoluene [(TNT) $(C_7H_5N_3O_6)$], first prepared by Wilbrand in 1863. The first detailed study of the preparation of 2,4,6-trinitrotoluene was by Beilstein and Kuhlberh in 1870, when they discovered the isomer 2,4,5-trinitrotoluene. Pure 2,4,6-trinitrotoluene was prepared in 1880 by Hepp and its structure established in 1883 by Claus and Becker. The manufacture of TNT began in Germany in 1891 and in 1899 aluminium was mixed with TNT to produce an explosive composition. In 1902, TNT was adopted for use by the German Army replacing picric acid, and in 1912 the US Army also started to use TNT. By 1914, TNT (1.4) became the standard explosive for all armies during World War I.

CH_3

O_2N NO_2

NO_2

(1.4)

Production of TNT was limited by the availability of toluene from coal tar and it failed to meet demand for the filling of munitions. Use of a mixture of TNT and ammonium nitrate, called amatol, became widespread to relieve the shortage of TNT. Underwater explosives used the same formulation with the addition of aluminium and was called aminal.

Development of Nitroguanidine

The explosive nitroguanidine was also used in World War I by the Germans as an ingredient for bursting charges. It was mixed with ammonium nitrate and paraffin for filling trench mortar shells. Nitroguanidine was also used during World War II and later in triple-base propellants.

Nitroguanidine ($CH_4N_4O_2$) was first prepared by Jousselin in 1877 and its properties investigated by Vieille in 1901. In World War I nitroguanidine was mixed with nitrocellulose and used as a flashless propellant. However, there were problems associated with this composition; nitroguanidine attacked nitrocellulose during its storage. This problem was overcome in 1937 by the company Dynamit AG who developed a propellant composition containing nitroguanidine called 'Gudol Pulver'. Gudol Pulver produced very little smoke, had no evidence of a muzzle flash on firing, and was also found to increase the life of the gun barrel.

After World War I, major research programmes were inaugurated to find new and more powerful explosive materials. From these programmes came cyclotrimethylenetrinitramine [(RDX) ($C_3H_6N_6O_6$)] also called Cyclonite or Hexogen, and pentaerythritol tetranitrate [(PETN) ($C_5H_8N_4O_{12}$)].

Development of PETN

PETN was first prepared in 1894 by nitration of pentaerythritol. Commercial production of PETN could not be achieved until formaldehyde and acetaldehyde required in the synthesis of pentaerythritol became readily available about a decade before World War II. During World War II, RDX was utilized more than PETN because PETN was more sensitive to impact and its chemical stability was poor. Explosive compositions containing 50% PETN and 50% TNT were developed and called 'Pentrolit' or 'Pentolite'. This composition was used for filling hand and anti-tank grenades, and detonators.

Development of RDX and HMX

RDX was first prepared in 1899 by the German, Henning for medicinal use. Its value as an explosive was not recognized until 1920 by Herz.

Herz succeeded in preparing RDX by direct nitration of hexamine, but the yields were low and the process was expensive and unattractive for large scale production. Hale, at Picatinny Arsenal in 1925, developed a process for manufacturing RDX which produced yields of 68%. However, no further substantial improvements were made in the manufacture of RDX until 1940 when Meissner developed a continuous method for the manufacture of RDX, and Ross and Schiessler from Canada developed a process which did not require the use of hexamine as a starting material. At the same time, Bachmann developed a manufacturing process for RDX (1.5) from hexamine which gave the greatest yield.

(1.5)

Bachmann's products were known as Type B RDX and contained a constant impurity level of 8–12%. The explosive properties of this impurity were later utilized and the explosive HMX, also known as Octogen, was developed. The Bachmann process was adopted in Canada during World War II, and later in the USA by the Tennessee–Eastman Company. This manufacturing process was more economical and also led to the discovery of several new explosives. A manufacturing route for the synthesis of pure RDX (no impurities) was developed by Brockman, and this became known as Type A RDX.

In Great Britain the Armament Research Department at Woolwich began developing a manufacturing route for RDX after the publication of Herz's patent in 1920. A small-scale pilot plant producing 75 lbs of RDX per day was installed in 1933 and operated until 1939. Another plant was installed in 1939 at Waltham Abbey and a full-scale plant was erected in 1941 near Bridgewater. RDX was not used as the main filling in British shells and bombs during World War II but was added to TNT to increase the power of the explosive compositions. RDX was used in explosive compositions in Germany, France, Italy, Japan, Russia, USA, Spain and Sweden.

Research and development continued throughout World War II to develop new and more powerful explosives and explosive compositions. Torpex (TNT/RDX/aluminium) and cyclotetramethylenetetranitramine, known as Octogen [(HMX)($C_4H_8N_8O_8$)], became available at

Table 1.1 *Examples of explosive compositions used in World War II*

Name	Composition
Baronal	Barium nitrate, TNT and aluminium
Composition A	88.3% RDX and 11.7% non-explosive plasticizer
Composition B (cyclotol)	RDX, TNT and wax
H-6	45% RDX, 30% TNT, 20% aluminium and 5% wax
Minol-2	40% TNT, 40% ammonium nitrate and 20% aluminium
Pentolites	50% PETN and 50% TNT
Picratol	52% Picric acid and 48% TNT
PIPE	81% PETN and 19% Gulf Crown E Oil
PTX-1	30% RDX, 50% tetryl and 20% TNT
PTX-2	41–44% RDX, 26–28% PETN and 28–33% TNT
PVA-4	90% RDX, 8% PVA and 2% dibutyl phthalate
RIPE	85% RDX and 15% Gulf Crown E Oil
Tetrytols	70% Tetryl and 30% TNT
Torpex	42% RDX, 40% TNT and 18% aluminium

the end of World War II. In 1952 an explosive composition called 'Octol' was developed; this contained 75% HMX and 25% TNT. Mouldable plastic explosives were also developed during World War II; these often contained vaseline or gelatinized liquid nitro compounds to give a plastic-like consistency. A summary of explosive compositions used in World War II is presented in Table 1.1.

Polymer Bonded Explosives

Polymer bonded explosives (PBXs) were developed to reduce the sensitivity of the newly-synthesized explosive crystals by embedding the explosive crystals in a rubber-like polymeric matrix. The first PBX composition was developed at the Los Alamos Scientific Laboratories in USA in 1952. The composition consisted of RDX crystals embedded in plasticized polystyrene. Since 1952, Lawrence Livermore Laboratories, the US Navy and many other organizations have developed a series of PBX formulations, some of which are listed in Table 1.2.

HMX-based PBXs were developed for projectiles and lunar seismic experiments during the 1960s and early 1970s using Teflon (polytetrafluoroethylene) as the binder. PBXs based on RDX and RDX/PETN have also been developed and are known as Semtex. Development is continuing in this area to produce PBXs which contain polymers that are energetic and will contribute to the explosive performance of the

Table 1.2 *Examples of PBX compositions, where HMX is cyclotetramethylene-tetranitramine (Octogen), HNS is hexanitrostilbene, PETN is pentaerythritol tetranitrate, RDX is cyclotrimethylenetrinitramine (Hexogen) and TATB is 1,3,5-triamino-2,4,6-trinitrobenzene*

Explosive	Binder and plasticizer
HMX	Acetyl-formyl-2,2-dinitropropanol (DNPAF) and polyurethane
HMX	Cariflex (thermoplastic elastomer)
HMX	Hydroxy-terminated polybutadiene (polyurethane)
HMX	Hydroxy-terminated polyester
HMX	Kraton (block copolymer of styrene and ethylene–butylene)
HMX	Nylon (polyamide)
HMX	Polyester resin–styrene
HMX	Polyethylene
HMX	Polyurethane
HMX	Poly(vinyl) alcohol
HMX	Poly(vinyl) butyral resin
HMX	Teflon (polytetrafluoroethylene)
HMX	Viton (fluoroelastomer)
HNS	Teflon (polytetrafluoroethylene)
NTO	Cariflex (block copolymer of butadiene–styrene)
NTO/HMX	Cariflex (block copolymer of butadiene–styrene)
NTO/HMX	Estane (polyester polyurethane copolymer)
NTO/HMX	Hytemp (thermoplastic elastomer)
PETN	Butyl rubber with acetyl tributylcitrate
PETN	Epoxy resin–diethylenetriamine
PETN	Kraton (block copolymer of styrene and ethylene–butylene)
PETN	Latex with bis-(2-ethylhexyl adipate)
PETN	Nylon (polyamide)
PETN	Polyester and styrene copolymer
PETN	Poly(ethyl acrylate) with dibutyl phthalate
PETN	Silicone rubber
PETN	Viton (fluoroelastomer)
PETN	Teflon (polytetrafluoroethylene)
RDX	Epoxy ether
RDX	Exon (polychlorotrifluoroethylene/vinylidine chloride)
RDX	Hydroxy-terminated polybutadiene (polyurethane)
RDX	Kel-F (polychlorotrifluoroethylene)
RDX	Nylon (polyamide)
RDX	Nylon and aluminium
RDX	Nitro-fluoroalkyl epoxides
RDX	Polyacrylate and paraffin
RDX	Polyamide resin
RDX	Polyisobutylene/Teflon (polytetrafluoroethylene)
RDX	Polyester
RDX	Polystyrene
RDX	Teflon (polytetrafluoroethylene)
TATB/HMX	Kraton (block copolymer of styrene and ethylene–butylene)

Table 1.3 *Examples of energetic polymers*

Common name	Chemical name	Structure
GLYN (monomer)	Glycidyl nitrate	$H_2C\overset{\displaystyle O}{-}CH-CH_2ONO_2$
polyGLYN	Poly(glycidyl nitrate)	$-\!\!\left[-CH_2-\underset{\displaystyle }{\overset{\displaystyle CH_2ONO_2}{CH}}-O-\right]_n\!\!-$
NIMMO (monomer)	3-Nitratomethyl-3-methyl oxetane	$H_3C,\ CH_2ONO_2$ on C; $H_2C,\ CH_2$; O (oxetane ring)
polyNIMMO	Poly(3-nitratomethyl-3-methyl oxetane)	$-\!\!\left[-O-CH_2-\underset{}{\overset{H_3C\ \ CH_2ONO_2}{C}}-CH_2-\right]_n\!\!-$
GAP	Glycidyl azide polymer	$-\!\!\left[-CH_2-\underset{}{\overset{CH_2N_3}{CH}}-O-\right]_n\!\!-$
AMMO (monomer)	3-Azidomethyl-3-methyl oxetane	$H_3C,\ CH_2N_3$ on C; $H_2C,\ CH_2$; O (oxetane ring)
PolyAMMO	Poly(3-azidomethyl-3-methyl oxetane)	$-\!\!\left[-O-CH_2-\underset{}{\overset{H_3C\ \ CH_2N_3}{C}}-CH_2-\right]_n\!\!-$
BAMO (monomer)	3,3-Bis-azidomethyl oxetane	$N_3H_2C,\ CH_2N_3$ on C; $H_2C,\ CH_2$; O (oxetane ring)
PolyBAMO	Poly(3,3-bis-azidomethyl oxetane)	$-\!\!\left[-O-CH_2-\underset{}{\overset{N_3H_2C\ \ CH_2N_3}{C}}-CH_2-\right]_n\!\!-$

Table 1.4 *Examples of energetic plasticizers*

Common name	Chemical name	Structure
NENAs	Alkyl nitratoethyl nitramines	$\overset{\displaystyle NO_2}{\underset{\displaystyle |}{}}$ R–N–CH$_2$–CH$_2$ONO$_2$
EGDN	Ethylene glycol dinitrate	O$_2$NOH$_2$C–CH$_2$ONO$_2$
MTN	Metriol trinitrate	CH$_2$ONO$_2$ H$_3$C–C–CH$_2$ONO$_2$ CH$_2$ONO$_2$
BTTN	Butane-1,2,4-triol trinitrate	ONO$_2$ O$_2$NOH$_2$C–CH–CH$_2$–CH$_2$ONO$_2$
K10	Mixture of di- and tri-nitroethylbenzene	
BDNPA/ BDNFA	Mixture of bis-dinitropropylacetal and bis-dinitropropylformal	

PBX. Inert prepolymers have been substituted by energetic prepolymers [(mainly hydroxy terminated polybutadiene (HTPB)] in explosive compositions, in order to increase the explosive performance, without compromising its vulnerability to accidental initiation. In the last ten years it has become apparent that PBXs containing inert or energetic binders are more sensitive to impact compared to traditional explosive compositions. The addition of a plasticizer has reduced the sensitivity of PBXs whilst improving its processability and mechanical properties. Energetic plasticizers have also been developed for PBXs. Examples of energetic polymers and energetic plasticizers under investigation are presented in Tables 1.3 and 1.4, respectively.

Recent Developments

Recent developments in explosives have seen the production of hexanitrostilbene [(HNS) ($C_{14}H_6N_6O_{12}$)] in 1966 by Shipp, and triaminotrinitrobenzene {(TATB) [(NH_2)$_3C_6$(NO_2)$_3$]} in 1978 by Adkins and Norris. Both of these materials are able to withstand relatively high temperatures compared with other explosives. TATB was first prepared in 1888 by Jackson and Wing, who also determined its solubility characteristics. In the 1950s, the USA Naval Ordnance Laboratories recognized TATB as a useful heat-resistant explosive, and successful small-scale preparations and synthetic routes for large-scale production were achieved to give high yields.

Nitro-1,2,4-triazole-3-one [(NTO) ($C_2H_2N_4O_2$)] is one of the new explosives with high energy and low sensitivity. It has a high heat of reaction and shows autocatalytic behaviour during thermal decomposition. NTO was first reported in 1905 from the nitration of 1,2,4-triazol-3-one. There was renewed interest in NTO in the late 1960s, but it wasn't until 1987 that Lee, Chapman and Coburn reported the explosive properties of NTO. NTO is now widely used in explosive formulations, PBXs, and gas generators for automobile inflatable airbag systems. The salt derivatives of NTO are also insensitive and are potential energetic ballistic additives for solid rocket propellants.

2,4,6,8,10,12-Hexanitrohexaazaisowurtzitane ($C_6H_6N_{12}O_{12}$) or HNIW, more commonly called CL-20 belongs to the family of high energy dense caged nitramines. CL-20 was first synthesized in 1987 by Arnie Nielsen, and is now being produced at SNPE in France in quantities of 50–100kg on an industrial pilot scale plant.

Nitrocubanes are probably the most powerful explosives with a predicted detonation velocity of $> 10,000$ m s^{-1}. Cubanes were first synthesised at the University of Chicago, USA by Eaton and Cole in 1964. The US Army Armament Research Development Centre (ARDEC) then funded development into the formation of octanitrocubane [(ONC) ($C_8N_8O_{16}$)] and heptanitrocubane [(HpNC) ($C_8N_7O_{14}$)]. ONC and HpNC were successfully synthesised in 1997 and 2000 respectively by Eaton and co-workers. The basic structure of ONC is a cubane molecule where all the hydrogens have been replaced by nitro groups (1.6). HpNC is denser than ONC and predicted to be a more powerful, shock-insensitive explosive.

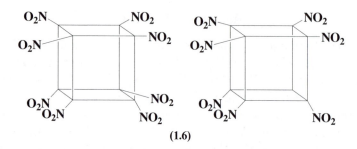

(1.6)

The research into energetic molecules which produce a large amount of gas per unit mass, led to molecular structures which have a high hydrogen to carbon ratio. Examples of these structures are hydrazinium nitroformate (HNF) and ammonium dinitramide (ADN). The majority of the development of HNF has been carried out in The Netherlands whereas the development of ADN has taken place in Russia, USA and Sweden. ADN is a dense non chlorine containing powerful oxidiser and is an interesting candidate for replacing ammonium perchlorate as an oxidiser for composite propellants. ADN is less sensitive to impact than RDX and HMX, but more sensitive to friction and electrostatic spark.

Insensitive Munitions

Recent developments of novel explosive materials have concentrated on reducing the sensitivity of the explosive materials to accidental initiation by shock, impact and thermal effects. The explosive materials, which have this reduced sensitivity, are call Insensitive Munitions, (IM). Although these explosive materials are insensitive to accidental initiation they still perform very well when suitably initiated. Examples of some explosive molecules under development are presented in Table 1.5. A summary of the significant discoveries in the history of explosives throughout the world is presented in Table 1.6.

Pollution Prevention

Historically waste explosive compositions (including propellants) have been disposed of by dumping the waste composition in the sea, or by burning or detonating the composition in an open bonfire. In 1994 the United Nations banned the dumping of explosive waste into the sea, and due to an increase in environmental awareness burning the explosive waste in an open bonfire will soon be banned since it is environmentally unacceptable. Methods are currently being developed to remove the waste explosive compositions safely from the casing using a high-

Table 1.5 *Examples of explosive molecules under development*

Common name	Chemical name	Structure
NTO	5-Nitro-1,2,4-triazol-3-one	
ADN	Ammonium dinitramide	
TNAZ	1,3,3-Trinitroazetidine	
CL-20	2,4,6,8,10,12-Hexanitro-2,4,6,8,10,12-hexa-azatetracyclododecane	
HNF	Hydrazinium nitroformate	$H_2N–NH_3C(NO_2)_3$
ONC	Octanitrocubane	
HpNC	Heptanitrocubane	
DAF	3,4-Diaminofurazan	

Table 1.6 *Some significant discoveries in the history of incendiaries, fireworks,
blackpowder and explosives*

Date	Explosive
220 BC	Chinese alchemists accidentally made blackpowder.
222–235 AD	Alexander VI of the Roman Empire called a ball of quicklime and asphalt 'automatic fire' which spontaneously ignited on coming into contact with water.
690	Arabs used blackpowder at the siege of Mecca.
940	The Chinese invented the 'Fire Ball' which is made of an explosive composition similar to blackpowder.
1040	The Chinese built a blackpowder plant in Pein King.
1169–1189	The Chinese started to make fireworks.
1249	Roger Bacon first made blackpowder in England.
1320	The German, Schwartz studied blackpowder and helped it to be introduced into central Europe.
1425	Corning, or granulating, process was developed.
1627	The Hungarian, Kaspar Weindl used blackpowder in blasting.
1646	Swedish Bofors Industries began to manufacture blackpowder.
1654	Preparation of ammonium nitrate was undertaken by Glauber.
1690	The German, Kunkel prepared mercury fulminate.
1742	Glauber prepared picric acid.
1830	Welter explored the use of picric acid in explosives.
1838	The Frenchman, Pelouze carried out nitration of paper and cotton.
1846	Schönbein and Böttger nitrated cellulose to produce guncotton.
1846	The Italian, Sobrero discovered liquid nitroglycerine.
1849	Reise and Millon reported that a mixture of charcoal and ammonium nitrate exploded on heating.
1863	The Swedish inventor, Nobel manufactured nitroglycerine.
1863	The German, Wilbrand prepared TNT.
1864	Schultze prepared nitrocellulose propellants.
1864	Nitrocellulose propellants were also prepared by Vieile.
1864	Nobel developed the mercury fulminate detonator.
1865	An increase in the stability of nitrocellulose was achieved by Abel.
1867	Nobel invented Dynamite.
1867	The Swedish chemists, Ohlsson and Norrbin added ammonium nitrate to dynamites.
1868	Brown discovered that dry, compressed guncotton could be detonated.
1868	Brown also found that wet, compressed nitrocellulose could be exploded by a small quantity of dry nitrocellulose.
1871	Abel proposed that ammonium picrate could be used as an explosive.
1873	Sprengel showed that picric acid could be detonated.
1875	Nobel mixed nitroglycerine with nitrocellulose to form a gel.

Continued

Table 1.6 *Continued*

Date	Explosive
1877	Mertens first prepared tetryl.
1879	Nobel manufactured Ammoniun Nitrate Gelatine Dynamite.
1880	The German, Hepp prepared pure 2,4,6-trinitrotoluene (TNT).
1883	The structure of tetryl was established by Romburgh.
1883	The structure of TNT was established by Claus and Becker.
1885	Turpin replaced blackpowder with picric acid.
1888	Jackson and Wing first prepared TATB.
1888	Picric acid was used in British Munitions called Liddite.
1888	Nobel invented Ballistite.
1889	The British scientists, Abel and Dewar patented Cordite.
1891	Manufacture of TNT began in Germany.
1894	The Russian, Panpushko prepared picric acid.
1894	Preparation of PETN was carried out in Germany.
1899	Preparation of RDX for medicinal use was achieved by Henning.
1899	Aluminium was mixed with TNT in Germany.
1900	Preparation of nitroguanidine was developed by Jousselin.
1902	The German Army replaced picric acid with TNT.
1905	NTO was first reported from the nitration of 1,2,4-triazol-3-one.
1906	Tetryl was used as an explosive.
1912	The US Army started to use TNT in munitions.
1920	Preparation of RDX by the German, Herz.
1925	Preparation of a large quantity of RDX in the USA.
1940	Meissner developed the continuous method for the manufacture of RDX.
1940	Bachmann developed the manufacturing process for RDX.
1943	Bachmann prepared HMX.
1952	PBXs were first prepared containing RDX, polystyrene and dioctyl phthalate in the USA.
1952	Octols were formulated.
1957	Slurry explosives were developed by the American, Cook.
1964	Cubanes were first synthesised at the University of Chicago, USA by Eaton and Cole.
1966	HNS was prepared by Shipp.
1970	The USA companies, Ireco and Dupont produced a gel explosive by adding paint-grade aluminium and MAN to ANFO.
1978	Adkins and Norris prepared TATB.
1983	TNAZ was first prepared at Fluorochem Inc.
1987	Lee, Chapman and Coburn reported the explosive properties of NTO.
1987	CL20 was first synthesized by Arnie Nielsen.
1997	ONC was successfully synthesised by Eaton and coworkers.
2000	HpNC was successfully synthesised by Eaton and coworkers.

pressure water jet. The recovered material then has to be disposed, one method is to reformulate the material into a commercial explosive. In the future, when formulating a new explosive composition, scientists must not only consider its overall performance but must make sure that it falls into the 'insensitive munitions' category and that it can easily be disposed or recycled in an environmentally friendly manner.

Chapter 2

Classification of Explosive Materials

EXPLOSIONS

An explosion occurs when a large amount of energy is suddenly released. This energy may come from an over-pressurized steam boiler, or from the products of a chemical reaction involving explosive materials, or from a nuclear reaction which is uncontrolled. In order for an explosion to occur there must be a local accumulation of energy at the site of the explosion which is suddenly released. This release of energy can be dissipated as blast waves, propulsion of debris, or by the emission of thermal and ionizing radiation.

These types of explosion can be divided into three groups; physical explosions such as the over-pressurized steam boiler, chemical explosions as in the chemical reactions of explosive compositions, and atomic explosions.

Atomic Explosions

The energy produced from an atomic or nuclear explosion is a million to a billion times greater than the energy produced from a chemical explosion. The shockwaves from an atomic explosion are similar to those produced by a chemical explosion but will last longer and have a higher pressure in the positive pulse and a lower pressure in the negative phase. The heavy flux of neutrons produced from an atomic explosion would be fatal to anybody near the explosion, whereas those who are some distance from the explosion would be harmed by the gamma radiation. Atomic explosions also emit intense infra-red and ultra-violet radiation.

Physical Explosions

A physical explosion can arise when a substance whilst being compressed undergoes a rapid physical transformation. At the same time, the potential energy of the substance is rapidly transformed into kinetic energy, and its temperature rises rapidly, resulting in the production of a shockwave in the surrounding medium.

An example of a physical explosion is the eruption of the Krakatoa volcano in 1883. During this eruption a large quantity of molten lava spilled into the ocean causing about 1 cubic mile of sea water to vapourize. This rapid vaporization created a blast wave which could by heard up to 3000 miles away.

Chemical Explosions

A chemical explosion is the result of a chemical reaction or change of state which occurs over an exceedingly short space of time with the generation of a large amount of heat and generally a large quantity of gas. Chemical explosions are produced by compositions which contain explosive compounds and which are compressed together but do not necessarily need to be confined. During a chemical explosion an extremely rapid exothermic transformation takes place resulting in the formation of very hot gases and vapours. Owing to the extreme rapidity of the reaction (one-hundredth of a second), the gases do not expand instantaneously but remain for a fraction of a second inside the container occupying the volume that was once occupied by the explosive charge. As this space is extremely small and the temperature of explosion is extremely high (several thousands of degrees), the resultant pressure is therefore very high (several hundreds of atmospheres) – high enough to produce a 'blast wave' which will break the walls of the container and cause damage to the surrounding objects. If the blast wave is strong enough, damage to distant objects can also occur.

The types of explosion described in this book are based on the explosions caused by the chemical reaction of explosive compositions.

CHEMICAL EXPLOSIVES

The majority of substances which are classed as chemical explosives generally contain oxygen, nitrogen and oxidizable elements (fuels) such as carbon and hydrogen. The oxygen is generally attached to nitrogen, as in the groups NO, NO_2 and NO_3. The exception to this rule are azides, such as lead azide (PbN_6), and nitrogen compounds such as nitrogen triiodide (NI_3) and azoimide (NH_3NI_3), which contain no oxygen.

In the event of a chemical reaction, the nitrogen and oxygen molecules separate and then unite with the fuel components as shown in Reaction 2.1.

$$
\begin{array}{c}
NO_2 \\
| \\
N \\
H_2C \diagup \diagdown CH_2 \\
| \qquad | \\
O_2N-N \diagdown \diagup N-NO_2 \\
CH_2
\end{array}
\longrightarrow \quad 3CO + 3H_2O + 3N_2
$$

(2.1)

During the reaction large quantities of energy are liberated, generally accompanied by the evolution of hot gases. The heat given out during the reaction (heat of reaction) is the difference between the heat required to break up the explosive molecule into its elements and the heat released on recombination of these elements to form CO_2, H_2O, N_2, *etc.*

Classification of Chemical Explosives

Classification of explosives has been undertaken by many scientists throughout this century, and explosives have been classified with respect to their chemical nature and to their performance and uses. Chemical explosives can be divided into two groups depending on their chemical nature; those that are classed as substances which are explosive, and those that are explosive mixtures such as blackpowder.

Substances that are explosive contain molecular groups which have explosive properties. Examples of these molecular groups are:

(1) nitro compounds;
(2) nitric esters;
(3) nitramines;
(4) derivatives of chloric and perchloric acids;
(5) azides;
(6) various compounds capable of producing an explosion, for example, fulminates, acetylides, nitrogen-rich compounds such as tetrazene, peroxides and ozonides, *etc.*

A systematic approach to the relationship between the explosive properties of a molecule and its structure was proposed by van't Hoff in 1909 and Plets in 1953. According to Plets, the explosive properties of any substance depend upon the presence of definite structural groupings. Plets divided explosives into eight classes as shown in Table 2.1.

Table 2.1 *Classification of explosive substances by their molecular groups*

Group	Explosive compounds
$-O-O-$ and $-O-O-O-$	Inorganic and organic peroxides and ozonides
$-OClO_2$ and $-OClO_3$	Inorganic and organic chlorates and perchlorates
$-N-X_2$	Where X is a halogen
NO_2 and $-ONO_2$	Inorganic and organic substances
$-N=N-$ and $-N=N=N-$	Inorganic and organic azides
$-N=C$	Fulminates
$-C\equiv C-$	Acetylene and metal acetylides
$M-C$	Metal bonded with carbon in some organometallic compounds

Classifying explosives by the presence of certain molecular groups does not give any information on the performance of the explosive. A far better way of classification is by performance and uses. Using this classification, explosives can be divided into three classes; (i) primary explosives, (ii) secondary explosives, and (iii) propellants as shown in Figure 2.1.

PRIMARY EXPLOSIVES

Primary explosives (also known as primary high explosives) differ from secondary explosives in that they undergo a very rapid transition from burning to detonation and have the ability to transmit the detonation to less sensitive explosives. Primary explosives will detonate when they are subjected to heat or shock. On detonation the molecules in the explosive dissociate and produce a tremendous amount of heat and/or shock. This will in turn initiate a second, more stable explosive. For these reasons, they are used in initiating devices. The reaction scheme for the decomposition of the primary explosive lead azide is given in Reaction 2.2.

$$\tfrac{1}{2}PbN_6 \rightarrow \tfrac{1}{2}Pb^{2+} + N_3^- \rightarrow \tfrac{1}{2}Pb + N_2 + N \qquad (2.2)$$

This reaction is endothermic, taking in 213 kJ of energy. According to Reaction 2.2, one atom of nitrogen is expelled from the N_3^- ion. This nitrogen then reacts with another N_3^- ion to form two molecules of nitrogen as shown in Reaction 2.3.

$$\tfrac{1}{2}PbN_6 + N \rightarrow \tfrac{1}{2}Pb^{2+} + N_3^- + N \rightarrow \tfrac{1}{2}Pb + 2N_2 \qquad (2.3)$$

Reaction 2.3 is highly exothermic, producing 657 kJ of energy. The decomposition of one N_3^- group may involve 2–3 neighbouring N_3^-

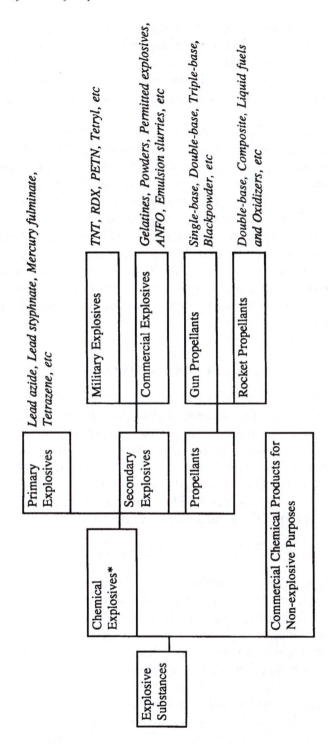

*Pyrotechnic compositions can also be classed as chemical explosives

Figure 2.1 *Classification of explosive substances*

groups. If these groups decompose simultaneously, the decomposition of 22 N_3^- ions may occur. Thus the rapid transition of lead azide to detonation may be accounted for by the fact that decomposition of a small number of molecules of lead azide may induce an explosion in a sufficiently large number of N_3^- ions to cause the explosion of the whole mass.

Primary explosives differ considerably in their sensitivity to heat and in the amount of heat they produce on detonation. The heat and shock on detonation can vary but is comparable to that for secondary explosives. Their detonation velocities are in the range of 3500–5500 m s^{-1}.

Primary explosives have a high degree of sensitivity to initiation through shock, friction, electric spark or high temperatures and explode whether they are confined or unconfined. Typical primary explosives which are widely used are lead azide, lead styphnate (trinitroresorcinate), lead mononitroresorcinate (LMNR), potassium dinitrobenzofurozan (KDNBF) and barium styphnate. Other primary explosive materials which are not frequently used today are mercury azide and mercury fulminate.

SECONDARY EXPLOSIVES

Secondary explosives (also known as high explosives) differ from primary explosives in that they cannot be detonated readily by heat or shock and are generally more powerful than primary explosives. Secondary explosives are less sensitive than primary explosives and can only be initiated to detonation by the shock produced by the explosion of a primary explosive. On initiation, the secondary explosive compositions dissociate almost instantaneously into other more stable components. An example of this is shown in Reaction 2.4.

$$C_3H_6N_6O_6 \rightarrow 3CO + 3H_2O + 3N_2 \qquad (2.4)$$

RDX ($C_3H_6N_6O_6$) will explode violently if stimulated with a primary explosive. The molecular structure breaks down on explosion leaving, momentarily, a disorganized mass of atoms. These immediately recombine to give predominantly gaseous products evolving a considerable amount of heat. The detonation is so fast that a shockwave is generated that acts on its surroundings with great brisance (or shattering effect) before the pressure of the exerted gas can take effect.

Some secondary explosives are so stable that rifle bullets can be fired through them or they can be set on fire without detonating. The more stable explosives which detonate at very high velocities exert a much greater force during their detonation than the explosive materials used to initiate them. Values of their detonation velocities are in the range of $5500–9000 \text{ m s}^{-1}$.

Examples of secondary explosives are TNT, tetryl, picric acid, nitrocellulose, nitroglycerine, nitroguanidine, RDX, HMX and TATB. Examples of commercial secondary explosives are blasting gelatine, guhr dynamite and 60% gelatine dynamite.

PROPELLANTS

Propellants are combustible materials containing within themselves all the oxygen needed for their combustion. Propellants only burn and do not explode; burning usually proceeds rather violently and is accompanied by a flame or sparks and a hissing or crackling sound, but not by a sharp, loud bang as in the case of detonating explosives. Propellants can be initiated by a flame or spark, and change from a solid to a gaseous state relatively slowly, *i.e.* in a matter of milliseconds. Examples of propellants are blackpowder, smokeless propellant, blasting explosives and ammonium nitrate explosives, which do not contain nitroglycerine or other aromatic nitro compounds.

CHEMICAL DATA ON EXPLOSIVE MATERIALS

Primary Explosives

Mercury Fulminate

Mercury fulminate ($C_2N_2O_2Hg$) (2.1) is obtained by treating a solution of mercuric nitrate with alcohol in nitric acid.

$$(C{\equiv}NO)_2Hg$$
$$(2.1)$$

This reaction, together with its products, has been studied by a number of chemists, including Liebig, who gave an account of the elementary chemical composition of fulminate in 1823. The mechanism of the reaction which results in the formation of mercury fulminate was reported by Wieland and by Solonina in 1909 and 1910, respectively.

The most important explosive property of mercury fulminate is that after initiation it will easily detonate. On detonation, it decomposes to stable products as shown in Reaction 2.5.

$$(CNO)_2Hg \rightarrow 2CO + N_2 + Hg \qquad (2.5)$$

Table 2.2 *Properties of mercury fulminate*

Characteristics	Explosive material
Colour	Grey, pale brown or white crystalline solid
Molecular weight	284.6
Decomposition temperature/°C	90
Thermal ignition temperature/°C	170
Crystal density at 20 °C/g cm^{-3}	4.42
Energy of formation/kJ kg^{-1}	+958
Enthalpy of formation/kJ kg^{-1}	+941

Mercury fulminate is sensitive to impact and friction, and is easily detonated by sparks and flames. It is desensitized by the addition of water but is very sensitive to sunlight and decomposes with the evolution of gas. Some of the properties of mercury fulminate are presented in Table 2.2.

Lead Azide

Lead azide (PbN$_6$) (2.2) was first prepared by Curtius in 1891.

$$N\!\!=\!\!N^+\!\!=\!\!N^-$$
$$Pb$$
$$N\!\!=\!\!N^+\!\!=\!\!N^-$$

(2.2)

Curtius added lead acetate to a solution of sodium or ammonium azide resulting in the formation of lead azide. In 1893, the Prussian Government carried out an investigation into using lead azide as an explosive in detonators, when a fatal accident occurred and stopped all experimental work in this area. No further work was carried out on lead azide until 1907 when Wöhler suggested that lead azide could replace mercury fulminate as a detonator. The manufacture of lead azide for military and commercial primary explosives did not commence until 1920 because of the hazardous nature of the pure crystalline material.

Lead azide has a good shelf life in dry conditions, but is unstable in the presence of moisture, oxidizing agents and ammonia. It is less sensitive to impact than mercury fulminate, but is more sensitive to friction. Lead azide is widely used in detonators because of its high capacity for initiating other secondary explosives to detonation. On a weight basis, it is superior to mercury fulminate in this role. However, since lead azide is

Table 2.3 *Properties of lead azide*

Characteristics	Explosive material
Colour	Colourless to white crystalline solid
Molecular weight	291.3
Decomposition temperature/°C	190
Thermal ignition temperature/°C	327–360
Energy of formation/kJ kg^{-1}	+1450
Enthalpy of formation/kJ kg^{-1}	+1420
Crystal density at 20 °C/g cm^{-3}	
α-form orthorhombic	4.17
β-form monoclinic	4.93

not particularly susceptible to initiation by impact it is not used alone in initiator components. Instead, it is used with lead styphnate and aluminium (ASA mixtures) for military detonators, in a mixture with tetrazene, and in a composite arrangement topped with a more sensitive composition.

Lead azide can exist in two allotropic forms; the more stable α-form which is orthorhombic, and the β-form which is monoclinic. The α-form is prepared by rapidly stirring a solution of sodium azide with a solution of lead acetate or lead nitrate, whereas the β-form is prepared by slow diffusion of sodium azide in lead nitrate solutions. The β-form has a tendency to revert to the α-form when its crystals are added to a solution containing either the α-form crystals or a lead salt. If the β-form crystals are left at a temperature of ~160 °C they will also convert to the α-form. Some of the properties of lead azide are presented in Table 2.3.

Lead Styphnate

Lead styphnate (2.3), also known as lead 2,4,6-trinitroresorcinate ($C_6H_3N_3O_9Pb$), is usually prepared by adding a solution of lead nitrate to magnesium styphnate. Lead styphnate is practically insoluble in water and most organic solvents. It is very stable at room- and elevated-temperatures (*e.g.* 75 °C) and is non-hygroscopic. Lead styphnate is exceptionally resistant towards nuclear radiation and can easily be ignited by a flame or electric spark. It is very sensitive to the discharge of static electricity and many accidents have occurred during its preparation. Attempts have been made to reduce its sensitiveness[1] to static

[1] In this book the term 'sensitiveness' is used to describe the response of an explosive to accidental initiation, *i.e.* friction, impact, *etc.*, whereas 'sensitivity' is used when the explosive is initiated by non-accidental methods, *i.e.* shock from a primary explosive.

Table 2.4 *Properties of lead styphnate (lead 2,4,6-trinitroresorcinate)*

Characteristics	Explosive material
Colour	Orange-yellow to dark brown crystalline solid
Molecular weight	468.3
Decomposition temperature/°C	235
Thermal ignition temperature/°C	267–268
Crystal density at 20 °C/g cm^{-3}	3.06
Energy of formation/kJ kg^{-1}	−1785
Enthalpy of formation/kJ kg^{-1}	−1826

discharge by adding graphite, but so far this has been unsuccessful and lead styphnate continues to be very dangerous to handle.

Lead styphnate is a weak primary explosive because of its high metal

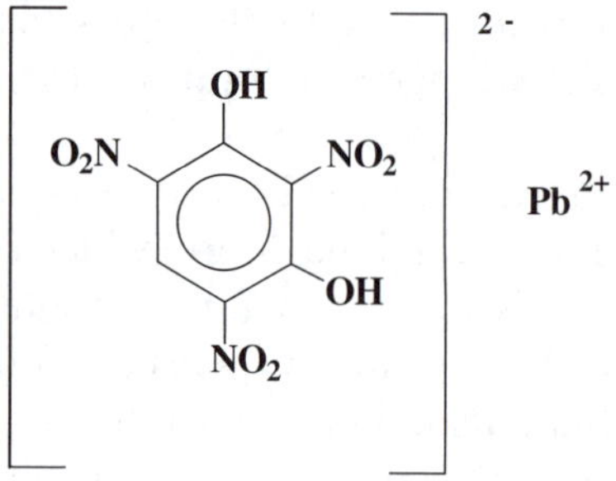

(2.3)

content (44.5%) and therefore is not used in the filling of detonators. It is used in ignition caps, and in the ASA (*i.e.* lead azide, lead styphnate and aluminium) mixtures for detonators. Some of its properties are shown in Table 2.4.

Silver Azide

Silver azide (AgN$_3$) (2.4) is manufactured in the same way as lead azide, by the action of sodium azide on silver nitrate in an aqueous solution.

$$AgN{=}N^+{=}N^-$$

(2.4)

Silver azide is slightly hygroscopic and is a very vigorous initiator, almost as efficient as lead azide. Like lead azide, silver azide decomposes under the influence of ultra-violet irradiation. If the intensity of radiation is sufficiently high the crystals may explode by photochemical

Table 2.5 *Properties of silver azide*

Characteristics	Explosive material
Colour	Fine white crystalline solid
Molecular weight	149.9
Melting temperature/°C	251
Thermal ignition temperature/°C	273
Crystal density at 20 °C/g cm^{-3}	5.1

decomposition. The ignition temperature and sensitiveness to impact of silver azide are lower than that of lead azide. Some of its properties are presented in Table 2.5.

Tetrazene

Tetrazene or tetrazolyl guanyltetrazene hydrate ($C_2H_8N_{10}O$) (2.5) was first prepared by Hoffmann and Roth in 1910 by the action of a neutral solution of sodium nitrite on aminoguanidine salts. In 1921, Rathsburg suggested the use of tetrazene in explosive compositions.

(2.5)

Tetrazene is slightly hygroscopic and stable at ambient temperatures. It hydrolyses in boiling water evolving nitrogen gas. Its ignition temperature is lower than that for mercury fulminate and it is slightly more sensitive to impact than mercury fulminate.

The detonation properties of tetrazene depend on the density of the material, *i.e.* its compaction. Tetrazene will detonate when it is not compacted but when pressed it produces a weaker detonation. These compaction properties make the transition from burning to detonation very difficult. Therefore, tetrazene is unsuitable for filling detonators. Tetrazene is used in ignition caps where a small amount is added to the explosive composition to improve its sensitivity to percussion and friction. Some of the properties of tetrazene are given in Table 2.6.

Table 2.6 *Properties of tetrazene*

Characteristics	Explosive material
Colour	Light colourless or pale yellow crystalline solid
Molecular weight	188.2
Decomposition temperature/°C	100
Thermal ignition temperature/°C	140
Crystal density at 20 °C/g cm^{-3}	1.7
Energy of formation/kJ kg^{-1}	+1130
Enthalpy of formation/kJ kg^{-1}	+1005

Secondary Explosives

Nitroglycerine

Nitroglycerine ($C_3H_5N_3O_9$) (2.6) was first prepared by the Italian, Ascanio Sobrero in 1846 by adding glycerol to a mixture of sulfuric and nitric acids. In 1863, a laboratory plant was set up to manufacture nitroglycerine by the Nobel family. In 1882, the Boutmy–Faucher process for the manufacture of nitroglycerine was developed in France and also adopted in England.

$$
\begin{array}{c}
H \\
| \\
H-C-O-NO_2 \\
| \\
H-C-O-NO_2 \\
| \\
H-C-O-NO_2 \\
| \\
H
\end{array}
$$

(2.6)

Nitroglycerine is a very powerful secondary explosive with a high brisance, *i.e.* shattering effect, and it is one of the most important and frequently-used components for gelatinous commercial explosives. Nitroglycerine also provides a source of high energy in propellant compositions, and in combination with nitrocellulose and stabilizers it is the principal component of explosive powders and solid rocket propellants.

Nitroglycerine is toxic to handle, causes headaches, and yields toxic products on detonation. It is insoluble in water but readily dissolves in most organic solvents and in a large number of aromatic nitro compounds, and forms a gel with nitrocellulose. The acid-free product is very stable, but exceedingly sensitive to impact. Some of the properties of nitroglycerine are presented in Table 2.7.

Table 2.7 *Properties of nitroglycerine*

Characteristics	Explosive material
Colour	Yellow oil
Molecular weight	227.1
Melting temperature/°C	13
Thermal ignition temperature/°C	200
Density at 20 °C/g cm^{-3}	1.59
Energy of formation/kJ kg^{-1}	-1547
Enthalpy of formation/kJ kg^{-1}	-1633

Nitrocellulose

Nitrocellulose (2.7) was discovered by C.F. Schönbeim at Basel and R. Böttger at Frankfurt-am-Main during 1845–47. The stability of nitro-cellulose was improved by Abel in 1865 and its detonation properties were developed by Abel's assistant, Brown.

(2.7)

The name, nitrocellulose does not refer to a single type of molecular compound but is a generic term denoting a family of compounds. The customary way to define its composition is to express the nitrogen content as a percentage by weight. The number of nitrate groups present in nitrocellulose can be calculated using Equation 2.1, where N is the percentage of nitrogen calculated from chemical analysis.

$$y = \frac{162N}{1400 - 45N} \tag{2.1}$$

If the structure for the unit cell of cellulose (2.8) is written as $C_6H_7O_2(OH)_3$, then nitrocellulose can be written as $C_6H_7O_2(OH)_x(ONO_2)_y$, where $x + y = 3$.

$$\begin{bmatrix} & \text{CH}_2\text{OH} \\ & | \\ & \text{C}\text{---}\text{O} \\ \text{H} \diagup \text{H} & \\ \text{C} & \text{C} \\ \diagdown & \\ \text{---O---} \quad \text{OH} \quad \text{H} \diagup \text{H} \\ \text{C}\text{---}\text{C} \\ \text{H} \quad \text{ÓH} \end{bmatrix}_n$$

(2.8)

A product containing on average two nitrate groups, *i.e.* $[\text{C}_6\text{H}_7\text{O}_2(\text{OH})_1(\text{ONO}_2)_2]$, will contain 11.11% nitrogen as shown in Equation 2.2.

$$2 = \frac{162N}{1400 - 45N}$$

$$2(1400 - 45N) = 162N$$

$$2800 - 90N = 162N$$

$$11.1 = N \qquad\qquad (2.2)$$

A product containing three nitrate groups, *i.e.* $[\text{C}_6\text{H}_7\text{O}_2(\text{ONO}_2)_3]$, will therefore contain 14.14% nitrogen. In practice, nitrocellulose compositions used in explosive applications vary from 10 to 13.5% of nitrogen.

Nitrocellulose materials prepared from cotton are fluffy white solids, which do not melt but ignite in the region of 180 °C. They are sensitive to initiation by percussion or electrostatic discharge and can be desensitized by the addition of water. The thermal stability of nitrocellulose decreases with increasing nitrogen content. The chemical stability of nitrocellulose depends on the removal of all traces of acid in the manufacturing process. Cellulose is insoluble in organic solvents, whereas nitrocellulose dissolves in organic solvents to form a gel. The gel has good physical properties due to the polymeric nature of nitrocellulose, and is an essential feature of gun propellants, double-base rocket propellants, gelatine and semi-gelatine commercial blasting explosives. Some of the properties of nitrocellulose are presented in Table 2.8.

Picric Acid

Glauber in 1742 reacted nitric acid with wool or horn and produced picric acid in the form of lead or potassium picrate. In 1771, Woulfe

Table 2.8 *Properties of nitrocellulose*

Characteristics	Explosive material
Colour	White fibres
Molecular weight of structural unit	324.2 + %N/14.14
Melting temperature/°C	Does not melt
Thermal ignition temperature/°C	190
Density at 20 °C/g cm^{-3}	1.67 max value by pressing
Energy of formation/kJ kg^{-1}	
13.3% Nitrogen	−2394
13.0% Nitrogen	−2469
12.5% Nitrogen	−2593
12.0% Nitrogen	−2719
11.5% Nitrogen	−2844
11.0% Nitrogen	−2999
Enthalpy of formation/kJ kg^{-1}	
13.3% Nitrogen	−2483
13.0% Nitrogen	−2563
12.5% Nitrogen	−2683
12.0% Nitrogen	−2811
11.5% Nitrogen	−2936
11.0% Nitrogen	−3094

prepared picric acid by treating indigo with nitric acid, and in 1778 Haussmann isolated picric acid also known as 2,4,6-trinitrophenol ($C_6H_3N_3O_7$) (2.9).

(2.9)

Other early experimenters obtained picric acid by nitrating various organic substances such as silk, natural resins, *etc.* The correct empirical formula for picric acid was determined by Laurent in 1841 who prepared the acid by reacting phenol with nitric acid and isolated dinitrophenol which was formed in an intermediate stage of the reaction.

Picric acid is a strong acid, very toxic, soluble in hot water, alcohol, ether, benzene and acetone, and is a fast yellow dye for silk and wool. It attacks common metals, except for aluminium and tin, and produces salts which are very explosive. The explosive power of picric acid is somewhat superior to that of TNT, both with regard to the strength and

Table 2.9 *Properties of picric acid (2,4,6-trinitrophenol)*

Characteristics	Explosive material
Colour	Yellow crystalline solid
Molecular weight	229.1
Melting temperature/$^{\circ}$C	122.5
Thermal ignition temperature/$^{\circ}$C	300
Crystal density at $20\,^{\circ}$C/g cm^{-3}	1.767
Energy of formation/kJ kg^{-1}	-873.8
Enthalpy of formation/kJ kg^{-1}	-944.3

the velocity of detonation.

Picric acid was used in grenade and mine fillings and had a tendency to form impact-sensitive metal salts (picrates) with the metal walls of the shells. The filling of mines and grenades was also a hazardous process, since relatively high temperatures were needed to melt the picric acid. Some of the properties of picric acid are presented in Table 2.9.

Tetryl

Tetryl (2.10), also known as 2,4,6-trinitrophenylmethylnitramine ($C_7H_5N_5O_8$), was first obtained by Mertens in 1877 and its structure determined by Romburgh in 1883.

(2.10)

Tetryl has been used as an explosive since 1906. In the early part of this century it was frequently used as the base charge of blasting caps but is now replaced by PETN or RDX. During World War II, it was used as a component of explosive mixtures.

Tetryl is a pale yellow, crystalline solid with a melting temperature of 129 $^{\circ}$C. It is moderately sensitive to initiation by friction and percussion, and is used in the form of pressed pellets as primers for explosive compositions which are less sensitive to initiation. It is slightly more sensitive than picric acid, and considerably more sensitive than TNT. Tetryl is quite toxic to handle and is now being replaced by RDX and

Table 2.10 *Properties of tetryl*

Characteristics	Explosive material
Colour	Light yellow crystalline solid
Molecular weight	287.1
Melting temperature/$^\circ$C	129.5
Decomposition temperature/$^\circ$C	130
Thermal ignition temperature/$^\circ$C	185
Crystal density at 20 $^\circ$C/g cm^{-3}	1.73
Energy of formation/kJ kg^{-1}	+195.5
Enthalpy of formation/kJ kg^{-1}	+117.7

wax (called Debrix). Some of the properties of tetryl are presented in Table 2.10.

TNT

Trinitrotoluene ($C_7H_5N_3O_6$) can exist as six different isomers. The isomer that is used in the explosives industry is the symmetrical isomer 2,4,6-trinitrotoluene (2.11). For convenience, the 2,4,6-isomer will be referred to in this book as TNT.

(2.11)

TNT was first prepared in 1863 by Wilbrand and its isomers discovered in 1870 by Beilstein and Kuhlberg. Pure TNT (2,4,6-trinitrotoluene isomer) was prepared by Hepp in 1880 and its structure determined by Claus and Becker in 1883. The development of TNT throughout the 19th and 20th centuries is summarized in Table 2.11.

TNT is almost insoluble in water, sparingly soluble in alcohol and will dissolve in benzene, toluene and acetone. It will darken in sunlight and is unstable in alkalis and amines. Some of the properties of TNT are presented in Table 2.12.

TNT has a number of advantages which have made it widely used in military explosives before World War I and up to the present time. These include low manufacturing costs and cheap raw materials, safety

Table 2.11 *Summary of the development of TNT (2,4,6-trinitrotoluene) throughout the 19th and 20th centuries*

Date	Developments
1837	Pelletier and Walter first prepared toluene.
1841	Mononitrated toluene was prepared.
1863	Wilbrand prepared crude TNT.
1870	Beilstein and Kuhlberg discovered the isomer 2,4,5-trinitrotoluene.
1880	Hepp prepared pure TNT (2,4,6-trinitrotoluene).
1882	The 2,3,4-trinitrotoluene isomer prepared.
1883	Claus and Becker determined structure of 2,4,6-trinitrotoluene.
1891	TNT was manufactured in Germany.
1899	Aluminium was added to TNT for use as explosives.
1900	Improved production of raw materials for TNT therefore reducing its cost.
1902	TNT replaced picric acid in the German Army.
1912	Use of TNT began in the US Army.
1913	Reduction in price of raw material (nitric acid) by the commercialization of the Haber–Bosch process for ammonia synthesis.
1914	Will determined structures of isomers 2,3,4- and 2,4,5-trinitrotoluene.
1914–18	Standard explosive for World War I. Production limited by availability of toluene from coal tar. To relieve shortage, TNT was mixed with ammonium nitrate to give amatols, and aluminium to produce tritonals.
1903–40	TNT was mixed with RDX to give cyclotols.
1939–45	During World War II, an improved process was developed for producing petroleum naphthas ensuring unlimited quantities of toluene. Purification techniques were improved for TNT. Composites mixtures of TNT–PETN, TNT–RDX, TNT–tetryl, TNT–ammonium picrate, TNT–aluminium, *etc.*, were prepared.
1952	TNT was mixed with HMX to give octols.
1966	Shipp prepared HNS from TNT.
1968	Continuous production of TNT in the USA.
1978	Adkins and Norris prepared TATB from TNT.

of handling, a low sensitivity to impact and friction, and a fairly high explosive power. TNT also has good chemical and thermal stability, low volatility and hygroscopicity, good compatibility with other explosives, a low melting point for casting, and moderate toxicity.

TNT is by far the most important explosive for blasting charges. It is widely used in commercial explosives and is much safer to produce and handle than nitroglycerine and picric acid. A lower grade of TNT can be used for commercial explosives, whereas the military grade is very pure.

Table 2.12 *Properties of TNT (2,4,6-trinitrotoluene)*

Characteristics	Explosive material
Colour	Pale yellow crystalline solid
Molecular weight	227.1
Melting temperature/$°C$	80.8
Thermal ignition temperature/$°C$	300
Crystal density at 20 $°C$/g cm^{-3}	1.654
Energy of formation/kJ kg^{-1}	-184.8
Enthalpy of formation/kJ kg^{-1}	-261.5

TNT can be loaded into shells by casting as well as pressing. It can be used on its own or by mixing with other components such as ammonium nitrate to give amatols, aluminium powder to give tritonal, RDX to give cyclonite and composition B.

One of the major disadvantages of TNT is the exudation (leaching out) of the isomers of dinitrotoluenes and trinitrotoluenes. Even a minute quantity of these substances can result in exudation. This often occurs in the storage of projectiles containing TNT, particularly in the summer time. The main disadvantage caused by exudation is the formation of cracks and cavities leading to premature detonation and a reduction in its density. Migration of the isomers to the screw thread of the fuse will form 'fire channels'. These fire channels can lead to accidental ignition of the charge. If the migrating products penetrate the detonating fuse, malfunctioning of the ammunition components can occur.

Nitroguanidine

Nitroguanidine (2.12), also known as picrite ($CH_4N_4O_2$), was first prepared by Jousselin in 1877 by dissolving dry guanidine nitrate in fuming nitric acid and passing nitrous oxide through the solution. The solution was poured into water and a precipitate was obtained which Jousselin called 'nitrosoguanidine', but which was later determined to be nitroguanidine by Thiele.

$$NH=C \begin{smallmatrix} NH_2 \\ \\ NH-NO_2 \end{smallmatrix}$$

(2.12)

Nitroguanidine is relatively stable below its melting point but decomposes immediately on melting to form ammonia, water vapour and solid

Table 2.13 *Properties of nitroguanidine*

Characteristics	Explosive material
Colour	White fibre-like crystalline solid
Molecular weight	104.1
Melting temperature/°C	
α-form	232
β-form	232
Thermal ignition temperature/°C	185
Decomposition temperature/°C	232
Crystal density at 20 °C/g cm^{-3}	1.71
Energy of formation/kJ kg^{-1}	−773.4
Enthalpy of formation/kJ kg^{-1}	−893.0

products. The gases from the decomposition of nitroguanidine are far less erosive than gases from other similar explosives. Nitroguanidine is soluble in hot water and alkalis, and insoluble in ethers and cold water. It has a high velocity of detonation, a low heat and temperature of explosion, and a high density. Some of the properties of nitroguanidine are given in Table 2.13.

Nitroguanidine can be used as a secondary explosive but is also suitable for use in flashless propellants as it possesses a low heat and temperature of explosion. These propellants contain a mixture of nitroguanidine, nitrocellulose, nitroglycerine and nitrodiethyleneglycol which together form a colloidal gel. Nitroguanidine does not dissolve in the gel, but becomes embedded as a fine dispersion. This type of colloidal propellant has the advantage of reducing the erosion of the gun barrel compared with conventional propellants.

PETN

PETN (2.13), also known as pentaerythritol tetranitrate ($C_5H_8N_4O_{12}$), is the most stable and the least reactive of the explosive nitric esters. It is insoluble in water, sparingly soluble in alcohol, ether and benzene, and soluble in acetone and methyl acetate. It shows no trace of decomposition when stored for a long time at 100 °C. It is relatively insensitive to friction but is very sensitive to initiation by a primary explosive.

$$O_2N-O-H_2C\diagdown_{\diagup}CH_2-O-NO_2$$
$$C$$
$$O_2N-O-H_2C\diagup^{\diagdown}CH_2-O-NO_2$$

(2.13)

Table 2.14 *Properties of PETN (pentaerythritol tetranitrate)*

Characteristics	Explosive material
Colour	Colourless crystalline solid
Molecular weight	316.1
Melting temperature/°C	141.3
Thermal ignition temperature/°C	202
Crystal density at 20 °C/g cm^{-3}	1.76
Energy of formation/kJ kg^{-1}	-1509
Enthalpy of formation/kJ kg^{-1}	-1683

PETN is a powerful secondary explosive and has a great shattering effect. It is used in commercial blasting caps, detonation cords and boosters. PETN is not used in its pure form because it is too sensitive to friction and impact. It is therefore usually mixed with plasticized nitrocellulose, or with synthetic rubbers to form polymer bonded explosives (PBXs). The most common form of explosive composition containing PETN is called Pentolite, which is a mixture of 20–50% PETN and TNT. PETN can also be formed into a rubber-like sheet and used in metal-forming, metal cladding and metal-hardening processes. PETN can also be incorporated into gelatinous industrial explosives.

In military applications PETN has been largely replaced by RDX since RDX is more thermally stable and has a longer shelf life. Some of the properties of PETN are given in Table 2.14.

RDX

RDX (2.14), also known as Hexogen, Cyclonite and cyclotrimethylene-trinitramine ($C_3H_6N_6O_6$), was first prepared in 1899 by Henning for medicinal use and used as an explosive in 1920 by Herz. The properties and preparation of RDX were fully developed during World War II.

$$NO_2$$
$$|$$
$$N$$
$$H_2C \diagup \quad \diagdown CH_2$$
$$O_2N-N \diagdown \quad \diagup N-NO_2$$
$$CH_2$$

(2.14)

RDX is a white, crystalline solid with a melting temperature of 204 °C. It attained military importance during World War II since it is more chemically and thermally stable than PETN and has a lower sensitiveness. Pure RDX is very sensitive to initiation by impact and friction and

Table 2.15 *Properties of RDX (cyclotrimethylenetrinitramine)*

Characteristics	Explosive material
Colour	White crystalline solid
Molecular weight	222.1
Melting temperature/°C	
Type A RDX	202–204
Type B RDX	192–193
Decomposition temperature/°C	213
Thermal ignition temperature/°C	260
Crystal density at 20 °C/g cm^{-3}	1.82
Energy of formation/kJ kg^{-1}	+417
Enthalpy of formation/kJ kg^{-1}	+318

is desensitized by coating the crystals with wax, oils or grease. It can also be compounded with mineral jelly and similar materials to give plastic explosives. Insensitive explosive compositions containing RDX can be achieved by embedding the RDX crystals in a polymeric matrix. This type of composition is known as a polymer bonded explosive (PBX) and is less sensitive to accidental initiation.

RDX has a high chemical stability and great explosive power compared with TNT and picric acid. It is difficult to dissolve RDX in organic liquids but it can be recrystallized from acetone. It has a high melting point which makes it difficult to use in casting. However, when it is mixed with TNT, which has a low melting temperature, a pourable mixture can be obtained. Some of the properties of RDX are presented in Table 2.15.

HMX

HMX (2.15), also known as Octogen and cyclotetramethylenetetra-nitramine ($C_4H_8N_8O_8$), is a white, crystalline substance which appears in four different crystalline forms differing from one another in their density and sensitiveness to impact. The β-form, which is least sensitive to impact, is employed in secondary explosives.

(2.15)

Table 2.16 *Properties of HMX (cyclotetramethylenetetranitramine)*

Characteristics	Explosive material
Colour	White crystalline solid
Molecular weight	296.2
Melting temperature/°C	275
Decomposition temperature/°C	280
Thermal ignition temperature/°C	335
Crystal density at 20 °C/g cm^{-3}	
α-form	1.87
β-form	1.96
γ-form	1.82
δ-form	1.78
Energy of formation/kJ kg^{-1}	+353.8
Enthalpy of formation/kJ kg^{-1}	+252.8

HMX is non-hygroscopic and insoluble in water. It behaves like RDX with respect to its chemical reactivity and solubility in organic solvents. However, HMX is more resistant to attack by sodium hydroxide and is more soluble in 55% nitric acid, and 2-nitropropane than RDX. In some instances, HMX needs to be separated from RDX and the reactions described above are employed for the separation. As an explosive, HMX is superior to RDX in that its ignition temperature is higher and its chemical stability is greater; however, the explosive power of HMX is somewhat less than RDX. Some of the properties of HMX are presented in Table 2.16.

TATB

TATB (2.16) is also known as 1,3,5-triamino-2,4,6-trinitrobenzene ($C_6H_6N_6O_6$) and was first obtained in 1888 by Jackson and Wing.

(2.16)

It is a yellow-brown coloured substance which decomposes rapidly just below its melting temperature. It has excellent thermal stability in the range 260–290 °C and is known as a heat-resistant explosive. Some of the properties of TATB are given in Table 2.17.

Table 2.17 *Properties of TATB (1,3,5-triamino-2,4,6-trinitrobenzene)*

Characteristics	Explosive material
Colour	Yellow-brown crystalline solid
Molecular weight	258.1
Melting temperature/°C	350
Decomposition temperature/°C	350
Thermal ignition temperature/°C	384
Crystal density at 20°C/g cm^{-3}	1.93
Energy of formation/kJ kg^{-1}	−425.0
Enthalpy of formation/kJ kg^{-1}	−597.9

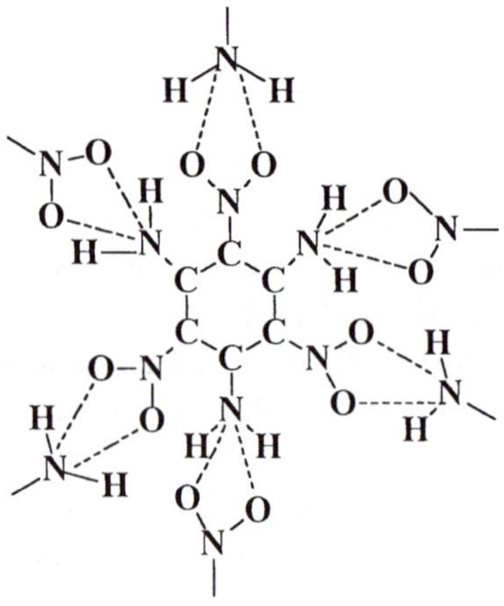

Figure 2.2 *Configuration of the TATB molecule*

The structure of TATB contains many unusual features. The unit cell of TATB consists of molecules arranged in planar sheets. These sheets are held together by strong intra- and intermolecular hydrogen bonding, resulting in a graphite-like lattice structure with lubricating and elastic properties as shown in Figure 2.2.

HNS

HNS [hexanitrostilbene (C$_{14}$H$_6$N$_6$O$_{12}$)] (2.17) is known as a heat-resistant explosive and is also resistant to radiation. It is practically insensitive to an electric spark and is less sensitive to impact than tetryl. Some of the properties of HNS are shown in Table 2.18.

Table 2.18 *Properties of HNS (hexanitrostilbene)*

Characteristics	Explosive material
Colour	Yellow crystalline solid
Molecular weight	450.1
Melting temperature/°C	318
Decomposition temperature/°C	318
Thermal ignition temperature/°C	325
Crystal density at 20 °C/g cm^{-3}	1.74
Energy of formation/kJ kg^{-1}	+195
Enthalpy of formation/kJ kg^{-1}	+128.1

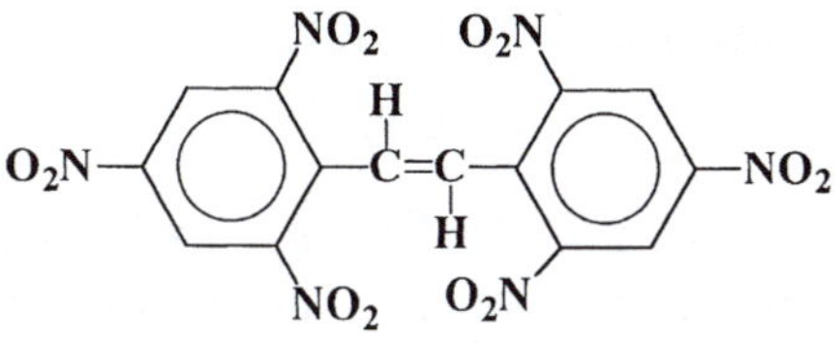

(2.17)

HNS is used in heat-resistant booster explosives and has been used in the stage separation in space rockets and for seismic experiments on the moon.

NTO

NTO [5-nitro-1,2,4-triazole-3-one (C$_2$H$_2$N$_4$O$_3$)] (2.18) is a new energetic material with attractive characteristics and high performance. It has a high heat of reaction and is less sensitive and more stable than RDX.

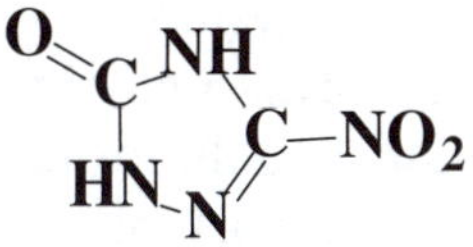

(2.18)

NTO is being developed in many areas these include i) a substitute for ammonium perchlorate or ammonium nitrate in solid rocket propellants, since it does not liberate undesirable products such as HCl and has quite a high burn rate compared to ammonium perchlorate and ammonium nitrate, ii) used as a burning rate modifier for composite propellants, iii) replacing RDX and HMX in composite solid propel-

Table 2.19 *Properties of NTO (5-nitro-1,2,4-triazole-3-one)*

Characteristics	Explosive material
Colour	White
Molecular weight	130.1
Decomposition temperature/°C	273
Thermal ignition temperature/°C	258–280
Crystal density at 20°C/g cm^{-3}	1.93
Enthalpy of formation/kJ kg^{-1}	−901
Heat of formation/kJ kg^{-1}	460–830

lants, and iv) improving the performance of gun propellants. Some of the properties of NTO are given in Table 2.19.

TNAZ

TNAZ (2.19) is also known as 1,3,3-trinitroazetidine ($C_3H_4N_4O_6$) and was first prepared in 1983 at Fluorochem Inc. It is reported to be more thermally stable than RDX but more reactive than HMX. Pure TNAZ is more shock sensitive than the explosives based on HMX but less sensitive than analogous PETN. It can be used as a castable explosive and as an ingredient in solid rocket and gun propellant.

(2.19)

TNAZ is a white crystalline solid and is soluble in acetone, methanol, ethanol, tetrachloromethane and cyclohexane. It has a relatively high vapour pressure and the solidification of its melt is accompanied by a high volume contraction resulting in the formation of shrinkage cavities giving 10–12% porosity. TNAZ exists in two crystal structures, the higher density crystal structure is more stable than the lower density crystal structure. The size of the crystals can be controlled by crystallizing TNAZ from a saturated solution of TNAZ and using an ultra-centrifuge, where the formation of defects in the crystal is minimised. An addition of 5–25% wt of a nitroaromatic amine to TNAZ can reduce the shock sensitivity of TNAZ. This is due to the additive acting as nucleation sites for TNAZ when it is crystallised from the melt, resulting in a more homogeneous structure. TNAZ forms eutectic mixtures with

Table 2.20 *Eutectic mixtures of TNAZ*

Additive	Melting point of additive/°C	Melting point of eutectic/°C	TNAZ content in eutectic/mol %
TNT	80.6	60.0	63.3–65.0
Tetryl	129.5	81.5–811.6	63.3–65.0
HMX	284.1	95.9	97.9

Table 2.21 *Properties of TNAZ (1,3,3-trinitroazetidine)*

Characteristics	Explosive material
Colour	White crystalline solid
Molecular weight	192.08
Melting temperature/°C	101
Boiling point/°C	252
Crystal density at 20°C/g cm^{-3}	1.84
Enthalpy of formation/kJ kg^{-1}	+190
Heat of melting/kJ mol^{-1}	28–30
Heat of sublimation/ kJ mol^{-1}	63.22
Heat of formation/ kJ mol^{-1}	936

some energetic materials, the details are presented in Table 2.20. Some properties of TNAZ are shown in Table 2.21.

OTHER COMPOUNDS USED IN EXPLOSIVE COMPOSITIONS

There are many other ingredients that are added to explosive compositions which in themselves are not explosive but can enhance the power of explosives, reduce the sensitivity, and aid processing. Aluminium powder is frequently added to explosive and propellant compositions to improve their efficiency. Ammonium nitrate (NH_4NO_3) is used extensively in commercial explosives and propellants. It is the most important raw material in the manufacture of commercial explosives and it also provides oxygen in rocket propellant compositions. Some of the properties of ammonium nitrate are presented in Table 2.22.

Commercial blasting explosives contain ammonium nitrate, wood meal, oil and TNT. A mixture of ammonium nitrate, water and oily fuels produces an emulsion slurry which is also used in commercial blasting explosives. Small glass or plastic spheres containing oxygen can be added to emulsion slurries to increase its sensitivity to detonation.

Polymeric materials can be added to secondary explosives to produce polymer bonded explosives (PBXs). The polymers are generally used in

Table 2.22 *Properties of ammonium nitrate*

Characteristics	Explosive material
Colour	Colourless crystalline solid
Molecular weight	80.0
Crystal range/°C	
α-form tetragonal	−18 to −16
β-form orthorhombic	−16 to 32.1
γ-form orthorhombic	32.1–84.2
δ-form orhombohedral or tetragonal	84.2–125.2
ε-form regular cubic (isometric)	125.2–169.6
Melting temperature/°C	169.6
Decomposition temperature/°C	169.6
Thermal ignition temperature/°C	169.6–210
Crystal density at 20 °C/g cm^{-3}	
α-form tetragonal	1.710
β-form orthorhombic	1.725
γ-form orthorhombic	1.661
δ-form orhombohedral or tetragonal	1.666
ε-form regular cubic (isometric)	1.594
Energy of formation/kJ kg^{-1}	−4424
Enthalpy of formation/kJ kg^{-1}	−4563

conjunction with compatible plasticizers to produce insensitive PBXs. The polymers and plasticizers can be in the nitrated form which will increase the power of the explosive. These nitrated forms are known as energetic polymers and energetic plasticizers.

Phlegmatizers are added to explosives to aid processing and reduce impact and friction sensitivity of highly sensitive explosives. Phlegmatizers can be waxes which lubricate the explosive crystals and act as a binder.

Chapter 3

Combustion, Deflagration and Detonation

When a loud, sharp bang is heard similar to a grenade or a bomb exploding it is known as detonation. If the noise is not as loud as that produced by a detonation and is longer in duration and sounds like a hissing sound (*i.e.* the sound of a rocket motor) it is classed as deflagration. In many cases these effects are preceded and accompanied by fire. If a fire is not accompanied by a thundering noise and 'blowing up' of a building, it is classed as either burning or combustion. Some explosive materials will burn relatively slowly (a few millimetres or centimetres per second) if spread on the ground in a thin line. The rate of burning will increase and sometimes develops into deflagration or detonation if these explosive materials are confined.

COMBUSTION

Combustion is a chemical reaction which takes place between a substance and oxygen. The chemical reaction is very fast and highly exothermic, and is usually accompanied by a flame. The energy generated during combustion will raise the temperature of the unreacted material and increase its rate of reaction. An example of this phenomenon can be seen when a matchstick is ignited. The initial process on striking a match is to create enough friction so that a large amount of heat is generated. This heat will locally raise the temperature of the match head so that the chemical reaction for combustion is initiated and the match head ignites. On ignition of the match head, heat is generated and the reactants burn in air with a flame. If the heat is reduced by wind blowing or by the wood of the matchstick being wet, the flame will extinguish.

Physical and Chemical Aspects of Combustion

Combustion is a complex process involving many steps which depend on the properties of the combustible substances. At low temperatures, oxidation of combustible materials can occur very slowly, without the presence of a flame. When the temperature is raised, as for example by the application of external heat, the rate of oxidation is increased. If the temperature of the reactants is raised to above their 'ignition temperature', the heat generated will be greater than the heat lost to the surrounding medium and a flame will be observed. Thus, when a lighted match is applied to butane gas the temperature of the gas is raised to the ignition point and a flame appears.

Combustion of Explosives and Propellants

The combustion process of propellant and explosive substances can be defined as a self-sustaining, exothermic, rapid-oxidizing reaction. Propellant and explosive substances will liberate a large amount of gas at high temperatures during combustion and will self-sustain the process without the presence of oxygen in the surrounding atmosphere. Propellants and explosives contain both oxidizer and fuel in their compositions and they are both classed as combustible materials. The chemical compositions of propellants and explosives are essentially the same; consequently some propellants can be used as explosives, and some explosives can be used as propellants. In general, propellants generate combustion gases by the deflagration process, whereas explosives generate these gases by deflagration or detonation. The combustion process of propellants is usually subsonic, whereas the combustion process of explosives during detonation is supersonic.

DEFLAGRATION

A substance is classed as a deflagrating explosive when a small amount of it in an unconfined condition suddenly ignites when subjected to a flame, spark, shock, friction or high temperatures. Deflagrating explosives burn faster and more violently than ordinary combustible materials. They burn with a flame or sparks, or a hissing or crackling noise.

On initiation of deflagrating explosives, local, finite 'hotspots' are developed either through friction between the solid particulates, by the compression of voids or bubbles in the liquid component, or by plastic flow of the material. This in turn produces heat and volatile intermediates which then undergo highly exothermic reactions in the gaseous phase. This whole process creates more than enough energy and heat to initiate the decomposition and volatilization of newly-exposed surfaces. Concomitantly, deflagration is a self-propagating process.

The rate of deflagration will increase with increasing degree of confinement. For example, a large pile of explosive material will contain particles that are confined. As the material undergoes deflagration the gases produced from the decomposition of the explosive crystals become trapped in the pile and thus raise the internal pressure. This in turn causes the temperature to rise resulting in an increase in the rate of deflagration.

The rate at which the surface of the composition burns, 'linear burning rate', can be calculated using Equation 3.1, where r is the linear burning rate in mm s^{-1}, P is the pressure at the surface of the composition at a given instant, β is the burning rate coefficient and α is the burning rate index.

$$r = \beta P^{\alpha} \tag{3.1}$$

The burning rate coefficient β depends upon the units of r and P, and the burning rate index α can be found experimentally by burning explosive substances at different pressures P and measuring the linear burning rate r. Values for α vary from 0.3 to greater than 1.0. The increase in the linear burning rate for a propellant when it is confined in a gun barrel can be calculated using Equation 3.1.

For example, if the linear rate of burning for a typical propellant at atmospheric pressure (9.869×10^{-2} N mm^{-2}) in an unconfined state is equal to 5 mm s^{-1} and the burning rate index is 0.528, then the value of β equals 16.98 mm s^{-1} (N mm^{-2})$^{1/0.528}$ as shown in Equation 3.2.

$$5 = \beta(9.869 \times 10^{-2})^{0.528}$$

$$\frac{5}{(9.869 \times 10^{-2})^{0.528}} = \beta$$

$$16.98 = \beta \tag{3.2}$$

On burning the propellant inside a gun barrel, the pressures increase by 4000 times and the linear burning rate is raised to 399 mm s^{-1} as shown in Equation 3.3:

$$r = 16.98 \times (4000 \times 9.869 \times 10^{-2})^{0.528}$$

$$r = 399 \text{ mm s}^{-1} \tag{3.3}$$

If a deflagrating explosive is initiated in a confined state (completely enclosed in a casing) the gaseous products will not be able to escape. The pressure will increase with consequent rapid increase in the rate of deflagration. If the rate of deflagration reaches a value of 1000–1800 m s^{-1} (1000–1800 × 10^3 mm s^{-1}) it becomes classed as a 'low order' detonation. If the rate increases to 5000 m s^{-1} (5000 × 10^3 mm s^{-1}) the detonation becomes 'high order'. Therefore, a given explosive may behave as a deflagrating explosive when unconfined, and as a detonating explosive when confined and suitably initiated.

The burning of a deflagrating explosive is therefore a surface phenomenon which is similar to other combustible materials, except that explosive materials do not need a supply of oxygen to sustain the burning. The amount of explosive material burning at the surface in a unit of time depends upon the surface area of the burning surface A, its density ρ and the rate at which it is burning r. The mass m of the explosive consumed in unit time can be calculated using Equation 3.4:

$$m = rA\rho \qquad\qquad (3.4)$$

The propagation of an explosion reaction through a deflagrating explosive is therefore based on thermal reactions. The explosive material surrounding the initial exploding site is warmed above its decomposition temperature causing it to explode. Explosives such as propellants exhibit this type of explosion mechanism. Transfer of energy by thermal means through a temperature difference is a relatively slow process and depends very much on external conditions such as ambient pressure. The speed of the explosion process in deflagrating explosives is always subsonic; that is, it is always less than the speed of sound.

DETONATION

Explosive substances which on initiation decompose via the passage of a shockwave rather than a thermal mechanism are called detonating explosives. The velocity of the shockwave in solid or liquid explosives is between 1500 and 9000 m s^{-1}, an order of magnitude higher than that for the deflagration process. The rate at which the material decomposes is governed by the speed at which the material will transmit the shockwave, not by the rate of heat transfer. Detonation can be achieved either by burning to detonation or by an initial shock.

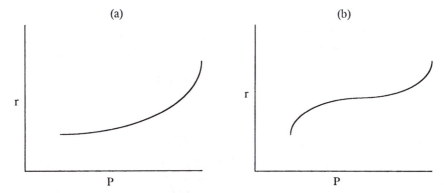

Figure 3.1 *Burning rate and pressure curves for detonating explosives, where (a) α increases to above unity, and (b) α increases further at higher pressures*

Burning to Detonation

Burning to detonation can take place when an explosive substance is confined in a tube and ignited at one end. The gas generated from the chemical decomposition of the explosive mixture becomes trapped, resulting in an increase in pressure at the burning surface; this in turn raises the linear burning rate. In detonating explosives the linear burning rate is raised so high by pressure pulses generated at the burning surface that it exceeds the velocity of sound, resulting in a detonation. The rise in the linear burning rate r with increasing pressure P for detonating explosives is shown in Figure 3.1. The values for r and P are derived from Equation 3.1.

The value for the burning rate index α is less than unity for deflagrating explosives. This value increases to above unity for detonating explosives [see Figure 3.1(a)] and may increase further at higher pressures as shown in Figure 3.1(b). An explosive which detonates in this way will show an appreciable delay between the initiation of burning and the onset of detonation as shown in Figure 3.2.

This delay will vary according to the nature of the explosive composition, its particle size, density and conditions of confinement. This principle of burning to detonation is utilized in delay fuses and blasting detonators.

Shock to Detonation

Explosive substances can also be detonated if they are subjected to a high velocity shockwave; this method is often used for the initiation of secondary explosives. Detonation of a primary explosive will produce a

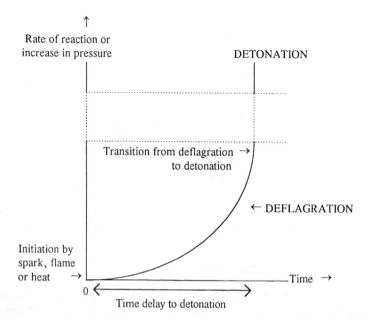

Figure 3.2 *Transition from deflagration to detonation*

shockwave which will initiate a secondary explosive if they are in close contact. The shockwave forces the particles to compress, and this gives rise to adiabatic heating which raises the temperature to above the decomposition temperature of the explosive material. The explosive crystals undergo an exothermic chemical decomposition which accelerates the shockwave. If the velocity of the shockwave in the explosive composition exceeds the velocity of sound, detonation will take place. Although initiation to detonation does not take place instantaneously the delay is negligible, being in microseconds.

Propagation of the Detonation Shockwave

The theory of detonation is a very complicated process containing many mathematical equations and far too complicated to be discussed here. The account given below is a very simplified qualitative version to give some basic understanding of the detonation process.

Suppose that a wave similar to a sound wave is produced in a column containing a gas by moving a piston inwards and outwards as shown in Figure 3.3.

This sound wave contains regions of rarefactions and compressions. The temperature of the material increases in the compression regions and then cools due to adiabatic expansion. In an explosive composition

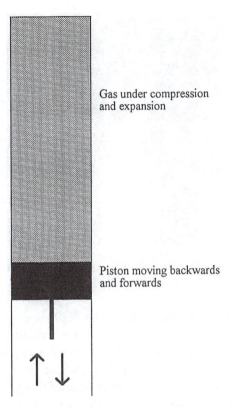

Gas under compression
and expansion

Piston moving backwards
and forwards

Figure 3.3 *Compression and expansion of a gas to produce a sound wave*

the compression part of the wave is sufficiently high to cause the temperature to rise above the decomposition temperature of the explosive crystals. As the explosive crystals decompose just behind the wave front, a large amount of heat and gas is generated. This in turn raises the internal pressure which contributes to the high pressures at the front of the wave. These high pressures at the wave front must be maintained for the wave front to move forward.

In order for the wave front to propagate forward (not laterally) and over a considerable distance, the explosive substance should either be confined inside a tube or have a cylindrical geometry. If the diameter of the explosive substance is too small, distortion of the wave front will occur, reducing its velocity and therefore causing the detonation wave to fade since the energy loss 'sideways' is too great for detonation to be supported. Consequently, the diameter of the explosive substance must be greater that a certain critical value, characteristic of the explosive substance.

Detonation along a cylindrical pellet of a secondary explosive can be regarded as a self-propagating process in which the axial compression of

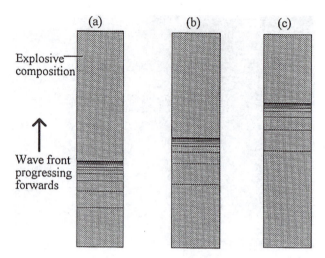

Figure 3.4 *Schematic diagram of the movement of a wave front through the explosive composition from (a) to (b) to (c)*

the shockwave changes the state of the explosive so that exothermic reactions take place. Figure 3.4 shows a schematic diagram for the progression of a wave front through a cylindrical explosive pellet.

The shockwave travels through the explosive composition accelerating all the time with increasing amplitude until it reaches a steady state. The conditions for a steady state are when the energy released from the chemical reactions equals (i) the energy lost to the surrounding medium as heat and (ii) the energy used to compress and displace the explosive crystals. At the steady state condition the velocity of a detonating wave will be supersonic.

On suitable initiation of a homogeneous liquid explosive, such as liquid nitroglycerine, the pressure, temperature, and density will all increase to form a detonation wave front. This will take place within a time interval of the order of magnitude of 10^{-12} s. Exothermic chemical reactions for the decomposition of liquid nitroglycerine will take place in the shockwave front. The shockwave will have an approximate thickness of 0.2 mm. Towards the end of the shockwave front the pressure will be about 220 kbar, the temperature will be above 3000 °C and the density of liquid nitroglycerine will be 30% higher than its original value.

Effect of Density on the Velocity of Detonation

For heterogeneous, commercial-type explosives the velocity of detonation increases and then decreases as the compaction density of the

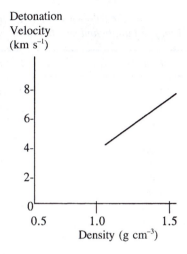

Figure 3.5 *Change in velocity of detonation as a function of density for a secondary explosive*, i.e. *TNT*

explosive composition increases. The compaction of heterogeneous explosives makes the transition from deflagration to detonation very difficult.

For homogeneous, military-type explosives the velocity of detonation will increase as the compaction density of the explosive composition increases as shown in Figure 3.5 and Table 3.1.

In order to achieve the maximum velocity of detonation for a homogeneous explosive, it is necessary to consolidate the explosive composition to its maximum density. For a crystalline explosive the density of compaction will depend upon the consolidation technique (*i.e.* pressing, casting, extrusion, *etc.*). The limiting density will be the density of the explosive crystal. The velocity of detonation can be calculated from the density of the explosive composition using Equation 3.5,

$$V_{\rho 1} = V_{\rho 2} + 3500\,(\rho_1 - \rho_2) \tag{3.5}$$

where $V_{\rho 1}$ and $V_{\rho 2}$ are the velocities of detonation for densities ρ_1 and ρ_2, respectively. The approximate velocity of detonation can be calculated using Equation 3.6,

$$V_{\rho x} = 430\,(nT_{\mathrm{d}})^{1/2} + 3500\,(\rho_x - 1) \tag{3.6}$$

where $V_{\rho x}$ is the velocity of detonation for a given density of compaction

Table 3.1 *The effect of density on the velocity of detonation for the primary*
explosive, mercury fulminate and secondary explosive, nitroguanidine

Explosive	Density/g cm^{-3}	Detonation velocity/m s^{-1}
Mercury fulminate	1.25	2300
	1.66	2760
	3.07	3925
	3.30	4480
	3.96	4740
Nitroguanidine	0.80	4695
	0.95	5520
	1.05	6150
	1.10	6440
	1.20	6775

Table 3.2 *A comparison of velocities of detonation for some primary and*
secondary explosives

Explosive	Density/g cm^{-3}	Detonation velocity/m s^{-1}
Primary explosives		
Lead styphnate	2.9	5200
Lead azide	3.8	4500
Mercury fulminate	3.3	4480
Secondary explosives		
HMX	1.89	9110
RDX	1.70	8440
PETN	1.60	7920
Picric acid	1.60	7900
Nitroguanidine	1.55	7650
TATB	1.88	7760
Nitroglycerine	1.60	7750
Nitrocellulose (dry)	1.15	7300
Tetryl	1.55	7080
HNS	1.70	7000
TNT	1.55	6850

ρ_x, n is the number of moles per gram of gaseous products produced from the detonation and T_d is the approximate temperature in Kelvin at which the detonation occurs.

A comparison of the velocities of detonation for some primary and secondary explosives is presented in Table 3.2.

From Table 3.2 it can be seen that the velocity of detonation for secondary explosives is generally higher than that for primary explosives.

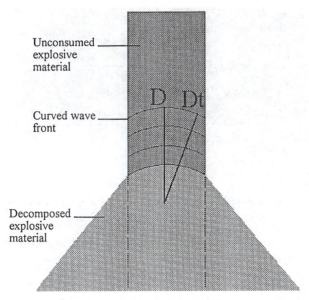

Unconsumed explosive material

Curved wave front

Decomposed explosive material

D Dt

Figure 3.6 *Propagation of a detonation wave illustrating the curved wave front*

Effect of Diameter of the Explosive Composition on the Velocity of Detonation

For a cylindrical pellet of an explosive composition the velocity of detonation will increase as the diameter of the explosive composition increases up to a limiting value. The detonation wave front for a cylindrical pellet at steady state conditions is not flat but convex as shown in Figure 3.6, where D is the axial detonation velocity and Dt is the detonation velocity close to the surface of the composition.

From Figure 3.6 it can be seen that the velocity of detonation gradually diminishes from the centre of the pellet to its surface. For large pellets the surface effects do not affect the velocity of detonation to the same degree as for small diameter pellets. There will be a finite value for the diameter of the pellet when the surface effects are so great that the wave front will no longer be stable – this is called the critical diameter. This phenomenon exists for homogeneous military-type explosives only. For heterogeneous commercial explosives the velocity of detonation increases with diameter. The reason for the difference in behaviour of homogeneous and heterogeneous explosive compositions is due to the mechanism of detonation. Homogeneous explosives rely on intramolecular reactions for the propagation of the shockwave, whereas detonation in heterogeneous explosives depends upon intermolecular

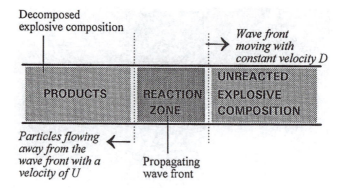

Figure 3.7 *Schematic diagram of a detonation wave moving forwards with the explosive particles moving in the opposite direction*

reactions which are diffusion-controlled since heterogeneous explosives need to be sensitized by air, bubbles, voids, *etc.*

Effect of Explosive Material on the Velocity of Detonation

The detonation process can therefore be regarded as a wave which is headed by a shock-front, which advances with constant velocity D into the unconsumed explosive, and is followed by a zone of chemical reaction as shown in Figure 3.7.

For the detonation wave to proceed forward its velocity in the reaction zone must equal the sum of the velocity of sound and the velocity of the flowing explosive material as shown in Equation 3.7,

$$D = U + c \qquad (3.7)$$

where D is the steady state velocity of the wave front, U is the velocity of the flowing particles and c is the velocity of a sound wave. When the velocity of the explosive particles is very low, *i.e.* U is low, the shockwave will be weak and its velocity will approach that of the speed of sound. Under these conditions a detonation will not take place. However, when the velocity of the explosive particles is high, *i.e.* U is high, the shockwave will travel faster than the speed of sound and a detonation will take place.

By applying the fundamental physical properties of conservation of mass, energy and momentum across the shockwave, together with the equation of state for the explosive composition (which describes the way its pressure, temperature, volume and composition affect one another) it can be shown that the velocity of detonation is determined by the material constituting the explosive and the material's velocity.

Table 3.3 *A comparison of effects for non-explosive combustible materials, deflagrating and detonating explosive materials*

	Non-explosive combustible substances	*Deflagrating explosive substances*	*Detonating explosive substances*
1	Initiated by flame, spark, high temps	Initiated by flame, sparks, friction, shock, high temps	Most explosives are capable of detonation if suitably initiated
2	Cannot be initiated in the wet state	Cannot be initiated in the wet state	Can be detonated in the wet state
3	Needs external supply of oxygen	Oxygen present in formulation	Oxygen present in formulation
4	Burns with a flame without any noise	Produces long, dull noise accompanied by hissing sound and fire	Loud, sharp bang, sometimes accompanied by fire
5	Burns with little generation of gases	Generation of gases used as propulsive forces in propellants	Generation of shockwave and used as a destructive force
6	Rate of burning slower than deflagration	Rate of burning is subsonic	Rate of burning is supersonic
7	Propagation based on thermal reactions	Propagation based on thermal reactions	Propagation based on shockwave
8	Rate of burning increases with increasing ambient pressure	Rate of burning increases with increasing ambient pressure	Velocity of detonation not affected by increasing ambient pressure
9	Not affected by strength of container	Not affected by strength of container	Velocity of detonation affected by strength of container
10	Not dependent on the size of the material	Not dependent on the size of the composition	Velocity of detonation dependent on diameter of explosive charge, *i.e.* critical diameter
11	Never converts to deflagration or detonation	Can convert to detonation if conditions are favourable	Does not usually revert to deflagration, if propagation of detonation wave fails explosive composition remains chemically unchanged

CLASSIFICATION OF EXPLOSIVES

Explosives can therefore be classified by the ease with which they can be ignited and subsequently exploded. Primary explosives are readily ignited or detonated by a small mechanical or electrical stimulus. Secondary explosives are not so easily initiated: they require a high velocity shockwave generally produced from the detonation of a primary explosive. Propellants are generally initiated by a flame, and they do not detonate, only deflagrate.

A comparison of effects for non-explosive combustible materials, deflagrating and detonating explosive materials is presented in Table 3.3.

Chapter 4

Ignition, Initiation and Thermal Decomposition

In most situations an event by a chemical explosive can be divided into four stages: these are ignition, the growth of deflagration, the transition from deflagration to detonation, and the propagation of detonation. In some circumstances ignition can lead straight to detonation. This only occurs when the initial stimulus is able to generate a large quantity of energy in the explosive composition. Heat is then produced by adiabatic compression in the shockwave front which results in detonation. Ignition to detonation only takes place in specially-formulated explosive compositions and requires special conditions and very high pressures.

IGNITION

Ignition occurs when part of a combustible material such as an explosive is heated to or above its ignition temperature. The ignition temperature is the minimum temperature required for the process of initiation to be self-sustaining.

Explosive materials are ignited by the action of an external stimulus which effectively inputs energy into the explosive and raises its temperature. The external stimulus can be friction, percussion, electrical impulse, heat, *etc.* Once stimulated the rise in temperature of the explosive causes a sequence of pre-ignition reactions to commence. These involve transitions in the crystalline structure, liquid phases changing into gaseous phases, and thermal decomposition of one or more of the ingredients. These reactions then lead to a self-sustaining combustion of the material, *i.e.* ignition. As the temperature rises, the rate of the heat produced increases exponentially whereas the rate of heat lost is linear. Ignition occurs at the temperature where the rate of heat generated is

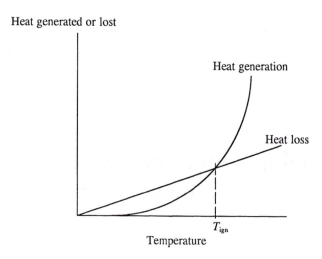

Figure 4.1 *Simple model to show ignition in explosives*

greater than the rate of heat lost. Figure 4.1 presents a simple model in defining the ignition temperature of an explosive material T_{ign}.

T_{ign} is the temperature at which the heat generated in the composition is greater than the heat lost to the surroundings, or more accurately, T_{ign} should equal 'ignition temperature − initial temperature'.

As discussed above, ignition generally results in deflagration of the explosive material, but if the material is confined or is in large quantities deflagration can develop into detonation. It is generally accepted that the initiation of explosives is a thermal process. Mechanical or electrical energy from the stimulus is converted into heat by a variety of mechanisms. The heat is concentrated in small regions forming hotspots.

Hotspots

The formation of hotspots depends upon the energy input and the physical properties of the explosive composition. The diameter of the hotspots is in the region of 0.1–10 μm and their duration is about 10^{-5}–10^{-3} s with temperatures greater than 900 °C. There have been various theories put forward to describe the mechanisms for the formation of hotspots, some of which are described below.

Mechanisms for the Formation of Hotspots

The energy from the stimulus is converted into heat by adiabatic compression of small, entrapped bubbles of gas. The heat generated forms hotspots. For an ideal gas, the temperature inside the gas bubbles T_2 depends on the compression ratio as shown in Equation 4.1,

$$T_2 = T_1 \left(\frac{P_2}{P_1} \right)^{\frac{\gamma - 1}{\gamma}} \tag{4.1}$$

where T_1 is the initial temperature of the bubble, P_1 and P_2 are the initial and final pressures inside the bubble, respectively, and γ is the ratio of the specific heats. From Equation 4.1 it can therefore be seen that when the initial pressure P_1 of the gas is raised, the temperature inside the bubble T_2 is reduced. The minimum temperature rise in the gas bubble must be about 450 °C for ignition to occur. This effect can be observed in the performance of liquid nitroglycerine when subjected to impact at different pressures. When liquid nitroglycerine is subjected to an impact energy of 5000 g cm^{-1} at an initial pressure of 1 atm an explosion is observed; however, if the initial pressure is then raised to 20–30 atm no explosive event takes place.

Under certain conditions these microscopic bubbles can result in an extremely sensitive explosive which can be ignited by the gentlest of blows. If precautions are taken to eliminate all the bubbles the explosive becomes comparatively insensitive, and very high impact energies must be used. Under these conditions the ignition is due to the viscous heating of the rapidly-flowing explosive as it escapes from between the impacting surfaces.

Another source of hotspots is the presence of grit particles, such as crystals. When the particles are small and sharp only a small amount of frictional or impact energy is needed to produce a hotspot. This is because localized energy is generated at the stress points; soft particles are unable to generate enough energy to produce hotspots since they will be crushed or squashed. Some explosive compositions which contain a polymer as a binder can also be quite sensitive to impact; this is due to the polymer failing catastrophically and releasing sufficient energy to form hotspots.

Ignition, therefore, begins at a hotspot but does not always lead to detonation. If the energy lost to the surroundings is greater than the energy generated by the hotspots the small micro-explosions die away without further propagation. If a particle of an explosive is smaller than a certain minimum size (which may be called 'critical'), the dissipation of heat will be greater than its evolution and no explosion will take place.

Table 4.1 *Temperatures for the initiation of some primary and secondary
explosives by friction (via hotspots) and thermal mechanisms*

Explosive substance	Temperature of ignition via hotspots/°C	Temperature of thermal ignition/°C
Primary explosives		
Tetrazene	~430	140
Mercury fulminate	~550	170
Lead styphnate	430–500	267
Lead azide	430–500	327–360
Secondary explosives		
Nitrogylcerine	450–480	200
PETN	400–430	202

IGNITION BY IMPACT AND FRICTION

Friction

When an explosive is subjected to friction, hotspots are formed. These
hotspots are generated by the rubbing together of explosive crystals that
are present in the explosive composition. Hotspots readily form on the
surface of the explosive crystals since they are non-metallic and have a
low thermal conductivity. The temperature of the hotspots must reach
temperatures greater than 430 °C for ignition of the explosive material
to occur. However, the maximum temperature of the hotspots is determined by the melting temperature of the crystals. Therefore, explosive
compositions containing crystals which have melting temperatures
lower than 430 °C will not achieve ignition through the formation of
hotspots, whereas those compositions with crystals of high melting
temperatures will form higher temperature hotspots capable of igniting
the material. Hotspots which are generated by friction are transient and
will only last for a very short time, *i.e.* 10^{-5} to 10^{-3} s. Consequently,
temperatures for ignition via hotspots is higher than conventional thermal ignition temperatures for explosive substances. Table 4.1 presents
data on the temperatures of ignition by friction via hotspots and thermal mechanisms for primary and secondary explosives.

Impact

When liquid explosives are subjected to high impact, compression and
heating of the trapped gases takes place and exothermic decomposition
of the explosive vapour begins. The rapid rise in temperature results in

further evaporation of the liquid explosive from the walls of the bubble, creating hotspots. These hotspots become sufficiently violent leading to the ignition of the liquid explosive. An increase in the quantity of gas bubbles will therefore result in an increase in the impact sensitivity of liquid and gelatinous explosive compositions.

When solid explosives are subjected to high impact, hotspots are formed from the compression and heating of the trapped gases, and from friction between the crystal particles. Ignition for the majority of primary explosives is via hotspots generated by intercrystalline friction, whereas ignition in secondary explosives is from hotspots generated through the compression of small gas spaces between the crystals. The difference between the formation of hotspots in primary and secondary explosives is related to the melting temperature of the crystals. Primary explosives will ignite below the melting temperature of the crystals whereas secondary explosives will ignite above the melting temperature.

When a secondary explosive is subjected to high impact, the material will flow (called 'plastic flow') like a liquid entrapping small gas bubbles. Hotspots will be generated by the compression and heating of the trapped gases similar to the process described for the formation of hotspots in liquid explosives, except that the impact energies needed for ignition will be far higher. Hotspots which are generated by impact are transient and will only last for a very short time in the order of 10^{-6} s. Consequently, hotspot temperatures for ignition are higher than conventional thermal ignition temperatures for explosive substances.

CLASSIFICATION OF EXPLOSIVES

Classification of substances by their sensitivity to impact and friction is particularly important for the handling of explosives. Some explosives are known to detonate on impact, whereas others will only deflagrate. Table 4.2 presents information on the sensitivity of explosive substances to impact and friction. The values shown describe the force required to initiate the explosive compositions.

The values given here clearly show that primary explosives are much more sensitive to friction than secondary explosives. Therefore, primary explosives are more hazardous to handle and care must be taken.

Another method of classifying high explosives is using the 'Figure of Insensitiveness' to impact and the 'Figure of Friction'. These values are obtained by subjecting powdered explosives either to impact using a Rotter Impact Machine, or by friction using a Rotary Friction testing machine.

In the Rotter impact test, the explosive samples are subjected to

Table 4.2 *Sensitivity of some primary and secondary explosives to impact and friction*

Substance	Friction sensitivity/N	Impact sensitivity/Nm
Primary explosives		
Tetrazene	8	1–2
Mercury fulminate	3–5	1–2
Lead styphnate	1.5	2.5–5
Lead azide	0.1–1	2.5–4
Secondary explosives		
PETN	60	3
HMX-β	120	7.4
RDX	120	7.5
HNS	240	5
Nitroglycerine	>353	0.2
Tetryl	>353	3
Nitrocellulose		
(13.4%N)	>353	3
Picric acid	>353	7.4
TNT	>353	15
Nitroguanidine	>353	>49
TATB	>353	50
Ammonium nitrate	>353	50

impact by a heavy weight, *i.e.* 5 kg, from different heights. The percentage of samples which ignite at a given height is noted. The results are plotted using the Bruceton Staircase technique and the median drop height which gives 50% probability of ignition for the materials under test is determined. The Figure of Insensitiveness (F of I) is calculated using Equation 4.2.

$$F \text{ of } I = \frac{\text{Median drop height of sample}}{\text{Median drop height of standard}} \times (F \text{ of } I \text{ of standard})$$

$$(4.2)$$

The Figure of Insensitiveness for the standard explosive sample 'RDX' is 80. Figure of Insensitiveness values for some primary and secondary explosives are presented in Table 4.3.

The Figure of Friction can be calculated in a similar way using the Bruceton Staircase technique, where the Figure of Friction for the standard explosive 'RDX' is 3.0. Figure of Friction (F of F) values for some secondary explosives are presented in Table 4.4.

Table 4.3 *Figure of Insensitiveness (F of I) for some primary and secondary high explosives calculated from the results of the Rotter Impact Machine (5 kg falling weight on to 30 mg samples)*

Explosive substance	Figure of Insensitiveness
Primary explosives	
Mercury fulminate	10
Tetrazene	13
Lead styphnate	12*
Lead azide	20*
Secondary explosives	
Nitrocellulose (dry 13.4% N)	23
Nitroglycerine	30
PETN	51
HMX-β	56
RDX	80
Gunpowder	90
Tetryl	86
TATB	> 100
Picric acid	120
TNT	152
Nitroguanidine	> 200

*2 kg weight falling on to 30 mg sample

Table 4.4 *Figure of Friction (F of F) for some secondary explosives calculated from the results of the Rotary Friction Machine*

Explosive substance	Figure of Friction
PETN	1.3
HMX-β	1.5
RDX	3.0
Tetryl	4.5
TNT	5.8

As with the Figure of Insensitiveness to impact, the lower the value for the Figure of Friction the more sensitive the material is.

Explosive substances can therefore be classified into three main groups using the results of the impact and friction sensitivity tests. These classes are 'very sensitive', 'sensitive' and 'comparatively insensitive'. By using the results of the tests carried out on the Rotter Impact Machine, explosive materials can be categorized into these three classifications as shown in Figure 4.2.

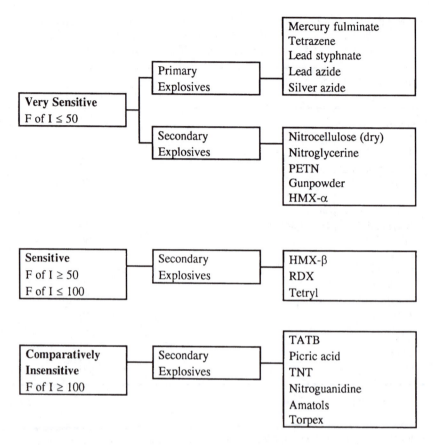

Figure 4.2 *Classification of primary and secondary explosives by their Figure of Insensitiveness (F of I)*

INITIATION TECHNIQUES

Explosive Train

An explosive composition is initiated or detonated via an explosive train. The explosive train is an arrangement of explosive components by which the initial force from the primer is transmitted and intensified until it reaches and sets off the main explosive composition. Some components of explosive trains are summarized in Table 4.5.

Almost all explosive trains contain a primary explosive as the first component. The second component in the train will depend upon the type of initiation process required for the main explosive composition. If this main explosive composition is to be detonated then the second component of the train will burn to detonation so that it imparts a shockwave to the main composition. This type of explosive train is

Table 4.5 *Some of the components of explosive trains*

Component	Action	Comments
Primer	Initiating device	Initiated by percussion, stabbing, electrical current, heat, *etc.*
Detonating	Detonate base charge	Ignited by primer. Small quantity of primary explosives
Flash	Ignite base charge	Ignited by primer. Burn explosively but will not detonate
Delay	Controlled time delay	Pyrotechnic formulation burns without gas
Relay	Initiate the next component	Its role is similar to the detonating component
Booster	Initiate main explosive composition	Used to initiate blasting agents or cast TNT
Base charge	Detonate main composition	Usually a secondary explosive

known as a 'detonator'. However, if the explosive train is only required to ignite the main composition an 'igniter' is used which will produce a flash instead of a detonation.

Detonators

Detonators are used for initiating explosives where a shockwave is required. Detonators can be initiated by electrical means, friction, flash from another igniferous element, stabbing and percussion. An example of an electrical detonator is presented in Figure 4.3.

In an electric detonator the bridgewire in the fusehead heats up to a temperature at which the sensitive composition surrounding the bridgewire ignites. This in turn will ignite the explosive composition in the priming charge. The priming charge will burn to detonation and send a shockwave into the base charge. This shockwave will initiate the explosive composition in the base charge to detonation. Detonators are used to explode secondary explosives, permissible explosives and dynamites.

Igniters

Igniters are used for initiating explosives whose nature is such that it is desirable to use a flame or flash for their initiation, and not a shock as produced by detonators. Explosives of this kind are known as deflagrat-

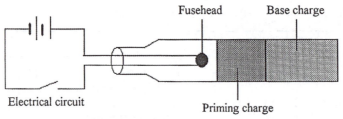

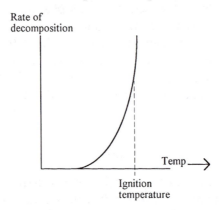

Figure 4.3 *Schematic diagram of an electric detonator*

ing explosives. Igniters can be initiated by electrical means, friction, flash from another igniferous element, stabbing and percussion. An example of an igniter is the 'squib', which is a small explosive device, similar in appearance to a detonator but loaded with an explosive which will deflagrate, so that its output is primarily heat (flash).

THERMAL DECOMPOSITION

All explosive substances undergo thermal decomposition at temperatures far below those at which explosions occur. During thermal decomposition, strong exothermic reactions take place which generate a lot of heat. Some of this heat is lost to the surroundings, but the remainder will raise the temperature of the explosives even further. When the rate of heat generated is greater than the rate of heat lost, spontaneous decomposition will occur, *i.e.* ignition. The increase in the rate of decomposition with temperature is shown in Figure 4.4.

Decomposition of explosive substances generally follows the curve given in Figure 4.4, where the rate rises very slowly at temperatures below 100 °C and then steepens as the temperature approaches the ignition temperature of the explosive. Explosives which have high igni-

Figure 4.4 *The effect of temperature on the rate of decomposition of an explosive*

tion temperatures, such as TATB, HMX, HNS and TNT, also have a high stability to heat, whereas those with low ignition temperatures have a low stability to heat.

The chemical energy H generated by the decomposition of the explosive can be calculated from the heat lost to the surroundings F and the accumulation of heat in the explosive Q as shown in Equation 4.3.

$$F + Q = H \qquad (4.3)$$

The amount of chemical energy H generated by the decomposition of an explosive will give information on the sensitivity of the explosive, since the mechanism for the initiation of explosives is thermal. Concomitantly, a high value for H will result in a more sensitive explosive.

Chapter 5

Thermochemistry of Explosives

Thermochemistry is an important part of explosive chemistry: it provides information on the type of chemical reactions, energy changes, mechanisms and kinetics which occur when a material undergoes an explosion. This chapter will carry out theoretical thermochemical calculations on explosive parameters, but it must be noted that the results obtained by such calculations will not always agree with those obtained experimentally, since experimental results will vary according to the conditions employed.

When an explosive reaction takes place, the explosive molecule breaks apart into its constituent atoms. This is quickly followed by a rearrangement of the atoms into a series of small, stable molecules. These molecules are usually water (H_2O), carbon dioxide (CO_2), carbon monoxide (CO) and nitrogen (N_2). There are also molecules of hydrogen (H_2), carbon (C), aluminium oxide (Al_2O_3), sulfur dioxide (SO_2), *etc.*, found in the products of some explosives. The nature of the products will depend upon the amount of oxygen available during the reaction. This supply of oxygen will depend in turn upon the quantity of oxidizing atoms which are present in the explosive molecule.

OXYGEN BALANCE

By considering the structural formula of TNT (5.1) and of nitroglycerine (5.2) the proportion of oxygen in each molecule can be calculated and compared with the amount of oxygen required for complete oxidation of the fuel elements, *i.e.* hydrogen and carbon.

If the amount of oxygen present in the explosive molecule is insufficient for the complete oxidation a negative oxygen balance will result; this can be seen in the molecule TNT. Nitroglycerine, however, has a

(5.1)   (5.2)

high proportion of oxygen, more than required for complete oxidation of its fuel elements and therefore has a positive oxygen balance. This oxygen balance can be defined as the amount of oxygen, expressed in weight percent, liberated as a result of the complete conversion of the explosive material to carbon dioxide, water, sulfur dioxide, aluminium oxide, *etc.*

When detonation of TNT ($C_7H_5N_3O_6$) takes place the explosive is oxidized to form gaseous products. Let us assume that on detonation the reactants are fully oxidized to form the gases carbon dioxide, water and nitrogen as shown in Reaction 5.1:

$$C_7H_5N_3O_6 \rightarrow nCO_2 + nH_2O + nN_2 \qquad (5.1)$$

The equation in Reaction 5.1 needs to be balanced, and this can be done by introducing oxygen atoms (O) as shown in Reaction 5.2.

$$C_7H_5N_3O_6 \rightarrow nCO_2 + nH_2O + nN_2 + nO$$

$$C_7H_5N_3O_6 \rightarrow 7CO_2 + nH_2O + nN_2 + nO$$

$$C_7H_5N_3O_6 \rightarrow 7CO_2 + 2\tfrac{1}{2}H_2O + nN_2 + nO$$

$$C_7H_5N_3O_6 \rightarrow 7CO_2 + 2\tfrac{1}{2}H_2O + 1\tfrac{1}{2}N_2 + nO$$

$$C_7H_5N_3O_6 \rightarrow 7CO_2 + 2\tfrac{1}{2}H_2O + 1\tfrac{1}{2}N_2 - 10\tfrac{1}{2}O \qquad (5.2)$$

In order to balance the reaction formula for the combustion of TNT a negative sign is used for oxygen. This therefore indicates that TNT has insufficient oxygen in its molecule to oxidize its reactants fully to form water and carbon dioxide. This amount of oxygen as percent by weight, can now be calculated as shown in Equation 5.1, where the atomic mass of carbon = 12, hydrogen = 1, nitrogen = 14 and oxygen = 16:

Formula of TNT ($C_7H_5N_3O_6$)

Molecular mass of TNT $= (7 \times 12) + (5 \times 1) + (3 \times 14) + (6 \times 16)$
$$= 227$$

Total molecular mass of oxygen atoms (O)
  in the products $= -10\frac{1}{2} \times 16 = -168$

Amount of oxygen liberated or taken in $= \dfrac{-168 \times 100}{227} = -74\%$

$$(5.1)$$

An alternative method for calculating the oxygen balance is shown in Equation 5.2. Here, the oxygen balance Ω is calculated from an explosive containing the general formula $C_aH_bN_cO_d$ with molecular mass M.

$$\Omega = \frac{[d - (2a) - (b/2)] \times 1600}{M} \qquad (5.2)$$

Using Equation 5.2 the oxygen balance for RDX ($C_3H_6N_6O_6$) is found to be -21.6% as shown in Equation 5.3:

For RDX, $a = 3$, $b = 6$, $c = 6$ and $d = 6$,

and $M = (3 \times 12) + (6 \times 1) + (6 \times 14) + (6 \times 16) = 222$

$$\Omega = \frac{[6 - (2 \times 3) - (6/2)] \times 1600}{222} = -21.6\% \qquad (5.3)$$

The balanced reaction formulae and calculated oxygen balances for some explosive substances are presented in Tables 5.1 and 5.2, respectively.

It can be seen from Table 5.2 that explosive substances may have a positive or negative oxygen balance. The oxygen balance provides information on the types of gases liberated. If the oxygen balance is large and negative then there is not enough oxygen for carbon dioxide to be formed. Consequently, toxic gases such as carbon monoxide will be liberated. This is very important for commercial explosives as the amount of toxic gases liberated must be kept to a minimum.

The oxygen balance does not provide information on the energy changes which take place during an explosion. This information can be obtained by calculating the heat liberated during decomposition of

Table 5.1 *Balanced reaction formulae for some explosives*

Explosive substance	Balanced reaction formulae for complete combustion
Ammonium nitrate	$NH_4NO_3 \rightarrow 2\,H_2O + N_2 + 1\,O$
Nitroglycerine	$C_3H_5N_3O_9 \rightarrow 3\,CO_2 + 2\frac{1}{2}\,H_2O + 1\frac{1}{2}\,N_2 + \frac{1}{2}\,O$
EGDN	$C_2H_4N_2O_6 \rightarrow 2\,CO_2 + 2\,H_2O + N_2 + 0\,O$
PETN	$C_5H_8N_4O_{12} \rightarrow 5\,CO_2 + 4\,H_2O + 2\,N_2 - 2\,O$
RDX	$C_3H_6N_6O_6 \rightarrow 3\,CO_2 + 3\,H_2O + 3\,N_2 - 3\,O$
HMX	$C_4H_8N_8O_8 \rightarrow 4\,CO_2 + 4\,H_2O + 4\,N_2 - 4\,O$
Nitroguanidine	$CH_4N_4O_2 \rightarrow CO_2 + 2\,H_2O + 2\,N_2 - 2\,O$
Picric acid	$C_6H_3N_3O_7 \rightarrow 6\,CO_2 + 1\frac{1}{2}\,H_2O + 1\frac{1}{2}\,N_2 - 6\frac{1}{2}\,O$
Tetryl	$C_7H_5N_5O_8 \rightarrow 7\,CO_2 + 2\frac{1}{2}\,H_2O + 2\frac{1}{2}\,N_2 - 8\frac{1}{2}\,O$
TATB	$C_6H_6N_6O_6 \rightarrow 6\,CO_2 + 3\,H_2O + 3\,N_2 - 9\,O$
HNS	$C_{14}H_6N_6O_{12} \rightarrow 14\,CO_2 + 3\,H_2O + 3\,N_2 - 19\,O$
TNT	$C_7H_5N_3O_6 \rightarrow 7\,CO_2 + 2\frac{1}{2}\,H_2O + 1\frac{1}{2}\,N_2 - 10\frac{1}{2}\,O$

Table 5.2 *Oxygen balance of some explosives*

Explosive substance	Empirical formula	Oxygen balance/% weight
Ammonium nitrate	NH_4NO_3	$+19.99$
Nitroglycerine	$C_3H_5N_3O_9$	$+3.50$
EGDN	$C_2H_4N_2O_6$	0.00
PETN	$C_5H_8N_4O_{12}$	-10.13
RDX	$C_3H_6N_6O_6$	-21.60
HMX	$C_4H_8N_8O_8$	-21.62
Nitroguanidine	$CH_4N_4O_2$	-30.70
Picric acid	$C_6H_3N_3O_7$	-45.40
Tetryl	$C_7H_5N_5O_8$	-47.39
TATB	$C_6H_6N_6O_6$	-55.80
HNS	$C_{14}H_6N_6O_{12}$	-67.60
TNT	$C_7H_5N_3O_6$	-74.00

explosive substances, known as the 'heat of explosion'. In order to calculate the heat of explosion, the decomposition products of the explosive must be determined, since the magnitude of the heat of explosion is dependent upon the thermodynamic state of its products. The decomposition process will be by detonation in the case of primary and secondary explosives, and burning in the case of gunpowders and propellants.

DECOMPOSITION REACTIONS

The detonation of HMX $(C_4H_8N_8O_8)$ will result in the formation of its decomposition products. These may be carbon monoxide, carbon dioxide, carbon, water, *etc.*, as shown in Reaction 5.3.

$$C_4H_8N_8O_8 \rightarrow 4CO + 4H_2O + 4N_2$$

or

$$C_4H_8N_8O_8 \rightarrow 2CO_2 + 2C + 4H_2O + 4N_2$$

or

$$C_4H_8N_8O_8 \rightarrow 2CO + 2CO_2 + 2H_2O + 2H_2 + 4N_2$$

or

$$C_4H_8N_8O_8 \rightarrow 3CO_2 + C + 2H_2O + 2H_2 + 4N_2$$

etc.   (5.3)

In order to clarify the problem of decomposition products, a set of rules was developed during World War II by Kistiakowsky and Wilson. These rules are nowadays known as the 'Kistiakowsky–Wilson' rules (K–W rules). These rules should only be used for moderately oxygen-deficient explosives with an oxygen balance greater than -40.0.

Kistiakowsky–Wilson Rules

The Kistiakowsky–Wilson rules are presented in Table 5.3.

Using these rules, the decomposition products of HMX ($C_4H_8N_8O_8$) can be determined as shown in Table 5.4.

Table 5.3 *Kistiakowsky–Wilson rules*

Rule no.	Conditions
1	Carbon atoms are converted to carbon monoxide
2	If any oxygen remains then hydrogen is then oxidized to water
3	If any oxygen still remains then carbon monoxide is oxidized to carbon dioxide
4	All the nitrogen is converted to nitrogen gas, N_2

Table 5.4 *Decomposition products of HMX using the Kistiakowsky–Wilson rules*

Rule no.	Conditions	Products
1	Carbon atoms are converted to carbon monoxide	$4C \rightarrow 4CO$
2	If any oxygen remains then hydrogen is then oxidized to water	$4O \rightarrow 4H_2O$
3	If any oxygen still remains then carbon monoxide is oxidized to carbon dioxide	No more oxygen
4	All the nitrogen is converted to nitrogen gas, N_2	$8N \rightarrow 4N_2$

Table 5.5 *Modified Kistiakowsky–Wilson rules*

Rule no.	Conditions
1	Hydrogen atoms are converted to water
2	If any oxygen remains then carbon is converted to carbon monoxide
3	If any oxygen still remains then carbon monoxide is oxidized to carbon dioxide
4	All the nitrogen is converted to nitrogen gas, N_2

Table 5.6 *Decomposition products of TNT using the modified Kistiakowsky–Wilson rules*

Rule no.	Conditions	Products
1	Hydrogen atoms are converted to water	$5\,H \rightarrow 2\frac{1}{2}\,H_2O$
2	If any oxygen remains then carbon is converted to carbon monoxide	$3\frac{1}{2}\,O \rightarrow 3\frac{1}{2}\,CO$
3	If any oxygen still remains then carbon monoxide is oxidized to carbon dioxide	No more oxygen
4	All the nitrogen is converted to nitrogen gas, N_2	$3\,N \rightarrow 1\frac{1}{2}\,N_2$

The overall reaction for the decomposition of HMX is given in Reaction 5.4.

$$C_4H_8N_8O_8 \rightarrow 4CO + 4H_2O + 4N_2 \qquad (5.4)$$

The Kistiakowsky–Wilson rules cannot be used for explosive materials which have an oxygen balance lower than -40. Under these circumstances the modified Kistiakowsky–Wilson rules must be employed.

Modified Kistiakowsky-Wilson Rules

The modified Kistiakowsky–Wilson rules (mod. K–W rules) are presented in Table 5.5.

Using these modified rules the decomposition products for TNT ($C_7H_5N_3O_6$) are given in Table 5.6.

The overall reaction for the decomposition of TNT is given in Reaction 5.5.

$$C_7H_5N_3O_6 \rightarrow 3\tfrac{1}{2}CO + 3\tfrac{1}{2}C + 2\tfrac{1}{2}H_2O + 1\tfrac{1}{2}N_2 \qquad (5.5)$$

The reaction for the decomposition of explosive substances can therefore be determined using either the Kistiakowsky–Wilson (K–W) rules

Table 5.7 *Reactions for decomposition products using the Kistiakowsky–Wilson (K–W) rules*

Explosive substance	Reaction for decomposition products
Nitroglycerine	$C_3H_5N_3O_9 \rightarrow 3\,CO_2 + 2\frac{1}{2}\,H_2O + \frac{1}{2}\,O + 1\frac{1}{2}\,N_2$
EGDN	$C_2H_4N_2O_6 \rightarrow 2\,CO_2 + 2\,H_2O + N_2$
PETN	$C_5H_8N_4O_{12} \rightarrow 2\,CO + 3\,CO_2 + 4\,H_2O + 2\,N_2$
RDX	$C_3H_6N_6O_6 \rightarrow 3\,CO + 3\,H_2O + 3\,N_2$
HMX	$C_4H_8N_8O_8 \rightarrow 4\,CO + 4\,H_2O + 4\,N_2$
Nitroguanidine	$CH_4N_4O_2 \rightarrow CO + H_2O + H_2 + 2\,N_2$

Table 5.8 *Reactions for decomposition products using the modified Kistiakowsky–Wilson (mod. K–W) rules*

Explosive substance	Reaction for decomposition products
Picric acid	$C_6H_3N_3O_7 \rightarrow 5\frac{1}{2}\,CO + \frac{1}{2}\,C + 1\frac{1}{2}\,H_2O + 1\frac{1}{2}\,N_2$
Tetryl	$C_7H_5N_5O_8 \rightarrow 5\frac{1}{2}\,CO + 1\frac{1}{2}\,C + 2\frac{1}{2}\,H_2O + 2\frac{1}{2}\,N_2$
TATB	$C_6H_6N_6O_6 \rightarrow 3\,CO + 3\,C + 3\,H_2O + 3\,N_2$
HNS	$C_{14}H_6N_6O_{12} \rightarrow 9\,CO + 5\,C + 3\,H_2O + 3\,N_2$
TNT	$C_7H_5N_3O_6 \rightarrow 3\frac{1}{2}\,CO + 3\frac{1}{2}\,C + 2\frac{1}{2}\,H_2O + 1\frac{1}{2}\,N_2$

or the modified Kistiakowsky–Wilson (mod. K–W) rules. Tables 5.7 and 5.8 present equations for the decomposition for some explosive materials using the Kistiakowsky–Wilson and the modified Kistiakowsky–Wilson rules, respectively.

A variation to Kistiakowsky–Wilson and modified Kistiakowsky–Wilson rules is provided by the Springall Roberts rules.

Springall Roberts Rules

The Springall Roberts rules take the unmodified Kistiakowsky–Wilson rules and add on two more conditions as shown in Table 5.9.

Using the Springall Roberts rules the products for the decomposition reaction of TNT ($C_7H_5N_3O_6$) are presented in Table 5.10.

The overall reaction for the decomposition of TNT is given in Reaction 5.6.

$$C_7H_5N_3O_6 \rightarrow 3CO + CO_2 + 3C + 1\frac{1}{2}H_2 + H_2O + 1\frac{1}{2}N_2 \qquad (5.6)$$

The three different rules for determining the decomposition products will only provide an insight to the decomposition products; they do not give any information as to the energy released on decomposition, but

Table 5.9 *Springall Roberts rules*

Rule no.	Conditions
1	Carbon atoms are converted to carbon monoxide
2	If any oxygen remains then hydrogen is then oxidized to water
3	If any oxygen still remains then carbon monoxide is oxidized to carbon dioxide
4	All the nitrogen is converted to nitrogen gas, N_2
5	One third of the carbon monoxide formed is converted to carbon and carbon dioxide
6	One sixth of the original amount of carbon monoxide is converted to form carbon and water

Table 5.10 *Decomposition products of TNT using the Springall Roberts rules*

Rule no.	Conditions	Products
1	Carbon atoms are converted to carbon monoxide	$6\,C \rightarrow 6\,CO$
2	If any oxygen remains then hydrogen is then oxidized to water	No more oxygen
3	If any oxygen still remains then carbon monoxide is oxidized to carbon dioxide	No more oxygen
4	All the nitrogen is converted to nitrogen gas, N_2	$3\,N \rightarrow 1\tfrac{1}{2}\,N_2$

This results in the formula,

$$C_7H_5N_3O_6 \rightarrow 6\,CO + C + 2\tfrac{1}{2}\,H_2 + 1\tfrac{1}{2}\,N_2$$

Rule no.	Conditions	Products
5	One third of the carbon monoxide formed is converted to carbon and carbon dioxide	$2\,CO \rightarrow C + CO_2$
6	One sixth of the original amount of carbon monoxide is converted to form carbon and water	$CO + H_2 \rightarrow C + H_2O$

are essential when calculating the heat of explosion. Although each rule will provide a different answer for the decomposition products they can be used as a guide. They are simple to apply and do give fairly good approximations.

HEATS OF FORMATION

The heats of formation for a reaction containing explosive chemicals can be described as the total heat evolved when a given quantity of a substance is completely oxidized in an excess amount of oxygen, resulting in the formation of carbon dioxide, water and sulfur dioxide. For explosive substances which do not contain sufficient oxygen in its molecule for complete oxidation, *i.e.* TNT, products such as carbon monoxide, carbon and hydrogen gas are formed. The energy liberated during the formation of these products is known as the 'heat of explosion'. If these products are then isolated and allowed to burn in excess oxygen to form substances like carbon dioxide, water, *etc.*, the heat evolved added to the heat of explosion would be equal to the 'heat of combustion'. Consequently, the value for the heat of combustion is higher than the value for the heat of explosion for substances which have insufficient oxygen for complete oxidation. For explosive substances with positive oxygen balances, *i.e.* nitroglycerine, there is generally no difference between the value for the heat of explosion and that of the heat of combustion.

The value for the heat of formation can be negative or positive. If the value is negative, heat is liberated during the reaction and the reaction is exothermic; whereas, if the value is positive, heat is absorbed during the reaction and the reaction is endothermic. For reactions involving explosive components the reaction is always exothermic. In an exothermic reaction the energy evolved may appear in many forms, but for practical purposes it is usually obtained in the form of heat. The energy liberated when explosives deflagrate is called the 'heat of deflagration', whereas the energy liberated by detonating explosives is called the 'heat of detonation' in kJ mol^{-1} or the 'heat of explosion' in kJ kg^{-1}.

In a chemical reaction involving explosives, energy is initially required to break the bonds of the explosive into its constituent elements as shown in Reaction 5.7 for RDX.

$$C_3H_6N_6O_6 \rightarrow 3C + 3H_2 + 3N_2 + 3O_2 \qquad (5.7)$$

These elements quickly form new bonds with the release of a greater quantity of energy as shown in Reaction 5.8.

$$3C + 3H_2 + 3N_2 + 3O_2 \rightarrow 3CO + 3H_2O + 3N_2 \qquad (5.8)$$

The molecules of an explosive are first raised to a higher energy level through input of the 'heats of atomization' in order to break their interatomic bonds. Then the atoms rearrange themselves into new molecules, releasing a larger quantity of heat and dropping to an energy level lower than the original as shown in Figure 5.1.

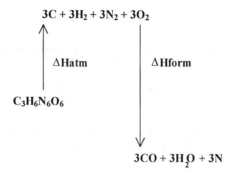

$$3C + 3H_2 + 3N_2 + 3O_2$$

ΔHatm

ΔHform

$$C_3H_6N_6O_6$$

$$3CO + 3H_2O + 3N$$

Figure 5.1 *Energy is taken in to break the bonds of RDX into its constituent elements, then energy is released when new bonds are formed*

The thermodynamic path presented in Figure 5.1 will most likely not be the same as the 'kinetic path'. For instance, the reaction may take place in several stages involving complex systems of reaction chains, *etc.* Nevertheless, the energy evolved depends only on the initial and final states and not on the intermediate ones. Once the reaction is completed, the net heat evolved is exactly the same as if the reactant molecules were first dissociated into their atoms, and then reacted directly to form the final products (Hess's Law). The heats of formation of some primary and secondary explosive substances are presented in Table 5.11.

HEAT OF EXPLOSION

When an explosive is initiated either to burning or detonation, its energy is released in the form of heat. The liberation of heat under adiabatic conditions is called the 'heat of explosion,' denoted by the letter Q. The heat of explosion provides information about the work capacity of the explosive, where the effective propellants and secondary explosives generally have high values of Q. For propellants burning in the chamber of a gun, and secondary explosives in detonating devices, the heat of explosion is conventionally expressed in terms of constant volume conditions Q_v. For rocket propellants burning in the combustion chamber of a rocket motor under conditions of free expansion to the atmosphere, it is conventional to employ constant pressure conditions. In this case, the heat of explosion is expressed as Q_p.

Consider an explosive which is initiated by a stimulus of negligible thermal proportions. The explosion can be represented by the irreversible process as shown in Figure 5.2, where Q is the value of the heat ultimately lost to the surroundings.

Table 5.11 *Heats of formation of some primary and secondary explosive substances*

Explosive substance	Empirical formula	Mol. wt.	ΔH_f /kJ kg^{-1}	ΔH_f /kJ mol^{-1}
Primary explosives				
Mercury fulminate	$HgC_2N_2O_2$	285	$+1354$	$+386$
Lead styphnate	$PbC_6H_3N_3O_9$	468	-1826	-855
Lead azide	PbN_6	291	$+1612$	$+469$
Secondary explosives				
Nitroglycerine	$C_3H_5N_3O_9$	227	-1674	-380
EGDN	$C_2H_4N_2O_6$	152	-1704	-259
PETN	$C_5H_8N_4O_{12}$	316	-1703	-538
RDX	$C_3H_6N_6O_6$	222	$+279$	$+62$
HMX	$C_4H_8N_8O_8$	296	$+253$	$+75$
Nitroguanidine	$CH_4N_4O_2$	104	-913	-95
Picric acid	$C_6H_3N_3O_7$	229	-978	-224
Tetryl	$C_7H_5N_5O_8$	287	$+118$	$+34$
TATB	$C_6H_6N_6O_6$	258	-597	-154
HNS	$C_{14}H_6N_6O_{12}$	450	$+128$	$+58$
TNT	$C_7H_5N_3O_6$	227	-115	-26

INITIATE

↓

EXPLOSION

↓

**GASEOUS PRODUCTS V
HEATS Q**

Figure 5.2 *Schematic diagram of the irreversible explosion process*

Under constant volume conditions Q_v can be calculated from the standard internal energies of formation for the products $\Delta U^\theta_{f\,(products)}$ and the standard internal energies of formation for the explosive components $\Delta U^\theta_{f\,(explosive\,components)}$ as shown in Equation 5.4.

$$Q_v = \Sigma\Delta U^\theta_{f\,(products)} - \Sigma\Delta U^\theta_{f\,(explosive\,components)} \qquad (5.4)$$

A similar expression is given for the heat of explosion under constant pressure conditions as shown in Equation 5.5, where ΔH^θ_f represents the corresponding standard enthalpies of formation:

$$Q_p = \Sigma \Delta H_f^\theta \text{(products)} - \Sigma \Delta H_f^\theta \text{(explosive components)} \tag{5.5}$$

In considering the thermochemistry of solid and liquid explosives, it is usually adequate, for practical purposes, to treat the state functions ΔH and ΔU as approximately the same. Consequently, heats, or enthalpy terms, tend to be used for both constant pressure and constant volume conditions.

Therefore, the heat of explosion Q can be calculated from the difference between the sum of the energies for the formation of the explosive components and the sum of the energies for the formation of the explosion products, as shown in Equation 5.6.

$$Q = \Delta H_{\text{(reaction)}} = \Sigma \Delta H_{f \text{ (products)}} - \Sigma \Delta H_{f \text{ (explosive components)}} \tag{5.6}$$

The calculated values do not exactly agree with those obtained experimentally since the conditions of loading density, temperature, pressure, *etc.*, are not taken into consideration.

The value for Q in kJ kg^{-1} is generally derived from the heat of detonation ΔH_d in kJ mol^{-1}. The heat of detonation for RDX can be illustrated using Hess's law as shown in Figure 5.3.

Using the diagram in Figure 5.3 the heat of detonation for RDX can be calculated as shown in Equation 5.7:

$$\Delta H_2 = \Delta H_1 + \Delta H_d \Rightarrow \Delta H_d = \Delta H_2 - \Delta H_1$$

Heat of formation of RDX $= \Delta H_f (\text{RDX}) = +62.0 \text{ kJ mol}^{-1}$

Heat of formation of carbon monoxide $= \Delta H_f (\text{CO})$
$$= -110.0 \text{ kJ mol}^{-1}$$

Heat of formation of water in the vapour phase $= \Delta H_f (\text{H}_2\text{O}_{(g)})$
$$= -242.0 \text{ kJ mol}^{-1}$$

Therefore, $\Delta H_1 = \Delta H_f (\text{RDX}) = +62.0 \text{ kJ mol}^{-1}$

$$\Delta H_2 = [3 \times \Delta H_f (\text{CO})] + [3 \times \Delta H_f (\text{H}_2\text{O}_{(g)})]$$
$$= [3 \times (-110.0)] + [3 \times (-242.0)] = -1056.0 \text{ kJ mol}^{-1}$$

$$\Delta H_d = \Delta H_2 - \Delta H_1 = -1056.0 - (+62.0) = -1118 \text{ kJ mol}^{-1} \tag{5.7}$$

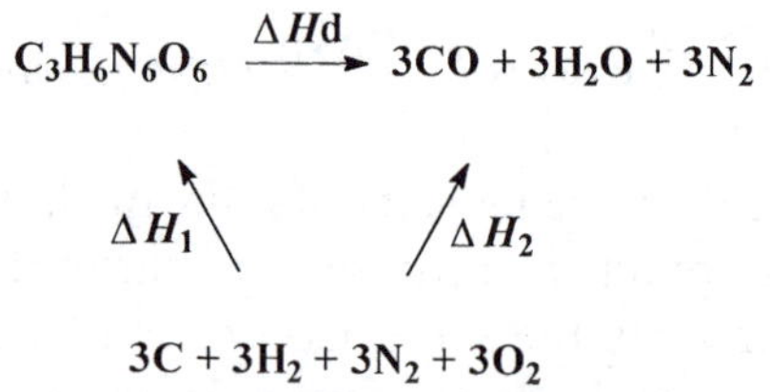

$$C_3H_6N_6O_6 \xrightarrow{\Delta Hd} 3CO + 3H_2O + 3N_2$$

$$\Delta H_1 \diagdown \qquad \diagup \Delta H_2$$

$$3C + 3H_2 + 3N_2 + 3O_2$$

Figure 5.3 *Enthalpy of detonation for RDX using Hess's law*

The heat of detonation for RDX is -1118 kJ mol^{-1}. This value can be converted to the heat of explosion Q as shown in Equation 5.8, where M is the molar mass of RDX (222):

$$Q = \frac{\Delta H_d \times 1000}{M} = \frac{-1118 \times 1000}{222} = -5036 \text{ kJ kg}^{-1} \qquad (5.8)$$

Another example is the calculation for the heat of explosion of PETN as shown in Figure 5.4 and in Equation 5.9:

$$\Delta H_d = \Delta H_2 - \Delta H_1$$

$$\Delta H_f(\text{PETN}) = -538.0 \text{ kJ mol}^{-1}$$

$$\Delta H_f(\text{CO}) = -110.0 \text{ kJ mol}^{-1}$$

$$\Delta H_f(\text{CO}_2) = -393.7 \text{ kJ mol}^{-1}$$

$$\Delta H_f(\text{H}_2\text{O}_{(g)}) = -242.0 \text{ kJ mol}^{-1}$$

Therefore, $\Delta H_1 = \Delta H_f(\text{PETN}) = -538.0 \text{ kJ mol}^{-1}$

$$\Delta H_2 = [2 \times \Delta H_f(\text{CO})] + [3 \times \Delta H_f(\text{CO}_2)] + [4 \times \Delta H_f(\text{H}_2\text{O}_{(g)})]$$

$$\Delta H_2 = [2 \times (-110.0)] + [3 \times (-393.7)] + [4 \times (-242.0)]$$

$$\Delta H_2 = -2369.1 \text{ kJ mol}^{-1}$$

$$\Delta H_d = \Delta H_2 - \Delta H_1 = -2369.1 - (-538) = -1831.1 \text{ kJ mol}^{-1}$$

$$Q = \frac{\Delta H_d \times 1000}{M} = \frac{-1831 \times 1000}{316} = -5794 \text{ kJ kg}^{-1} \qquad (5.9)$$

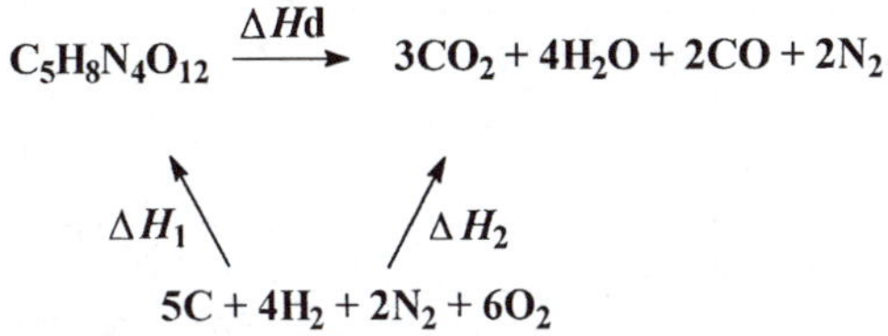

$$C_5H_8N_4O_{12} \xrightarrow{\Delta Hd} 3CO_2 + 4H_2O + 2CO + 2N_2$$

$$\Delta H_1 \diagdown \qquad \diagup \Delta H_2$$

$$5C + 4H_2 + 2N_2 + 6O_2$$

Figure 5.4 *Enthalpy of detonation for PETN using Hess's law*

Table 5.12 *Heat of explosion and heat of detonation at constant volume for some primary and secondary explosive substances using the K–W and modified K–W rules. $\Delta H_f(H_2O)$ is in the gaseous state*

Explosive substance	Empirical formula	ΔH_d/kJ mol^{-1}	Q_v/kJ kg^{-1}
Primary explosives			
Mercury fulminate	$HgC_2N_2O_2$	-500	1755
Lead styphnate	$PbC_6H_3N_3O_9$	-868	1855
Lead azide	PbN_6	-469	1610
Secondary explosives			
Nitroglycerine	$C_3H_5N_3O_9$	-1406	6194
EGDN	$C_2H_4N_2O_6$	-1012	6658
PETN	$C_5H_8N_4O_{12}$	-1831	5794
RDX	$C_3H_6N_6O_6$	-1118	5036
HMX	$C_4H_8N_8O_8$	-1483	5010
Nitroguanidine	$CH_4N_4O_2$	-257	2471
Picric acid	$C_6H_3N_3O_7$	-744	3249
Tetryl	$C_7H_5N_5O_8$	-1244	4335
TATB	$C_6H_6N_6O_6$	-902	3496
HNS	$C_{14}H_6N_6O_{12}$	-1774	3942
TNT	$C_7H_5N_3O_6$	-964	4247

The higher the value of Q for an explosive, the more the heat generated when an explosion occurs. Table 5.12 presents calculated heats of explosion together with their calculated heats of detonation for some primary and secondary explosive substances. The negative sign for the heat of explosion is generally omitted since it only denotes an exothermic reaction.

From Table 5.12 it can be seen that secondary explosives generate far more heat during an explosion than primary explosives.

Effect of Oxygen Balance

The effect of the oxygen balance on the heat of explosion can be seen from Figure 5.5.

The heat of explosion reaches a maximum for an oxygen balance of

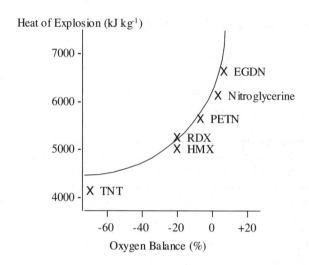

Heat of Explosion (kJ kg⁻¹)

Figure 5.5 *The effect of the oxygen balance on the heat of explosion*

zero, since this corresponds to the stoichiometric oxidation of carbon to
carbon dioxide and hydrogen to water. The oxygen balance can therefore be used to optimize the composition of the explosive to give an
oxygen balance as close to zero as possible. For example, TNT has an
oxygen balance of -74.0; it is therefore very deficient in oxygen, and by
mixing it with 79% ammonium nitrate which has an oxygen balance of
$+19.99$ the oxygen balance is reduced to zero resulting in a high value
for the heat of explosion.

VOLUME OF GASEOUS PRODUCTS OF EXPLOSION

The volume of gas produced during an explosion will provide information on the amount of work done by the explosive. In order to measure
the volume of gas generated standard conditions must be established,
because the volume of gas will vary according to the temperature at
which the measurement is taken. These standard conditions also enable
comparisons to be made between one explosive and another. The standard conditions set the temperature at $0\,°C$ or 273 K, and the pressure at
1 atm. These conditions are known as 'standard, temperature and
pressure', 'stp'. Under these standard conditions one mole of gas will
occupy 22.4 dm³, which is known as the molar gas volume. The volume
of gas V produced from an explosive during detonation can be calculated from its equation of decomposition, where information can be
obtained on the amount of gaseous products liberated. Examples for the
calculation of V during detonation of RDX and TNT are given below.

The equation for explosion of RDX using the K–W rules is given in Reaction 5.9.

$$C_3H_6N_6O_6 \rightarrow 3CO + 3H_2O + 3N_2 \qquad (5.9)$$

The production of water will be turned into steam as the temperature of explosion will be very high; therefore, water will be regarded as a gaseous product. From Reaction 5.9 it can be seen that 9 moles of gas are produced from 1 mole of RDX. Therefore, 9 moles of gas will occupy 201.6 dm^3 and 1 g of RDX will produce 0.908 dm^3 g^{-1} (908 cm^3 g^{-1}) of gas at 'stp' as shown in Equation 5.10:

1 mol of gas at 'stp' will occupy 22.4 dm^3

9 mol of gas will occupy 22.4 × 9 = 201.6 dm^3

Therefore, 1 mol of RDX produces 201.6 dm^3 of gas

1 g of RDX produces $\dfrac{201.6}{222}$ = 0.908 dm^3 g^{-1}

where the molecular mass of RDX is 222 $\qquad (5.10)$

The equation for explosion of TNT using the modified K–W rules is shown in Reaction 5.10.

$$C_7H_5N_3O_6 \rightarrow 3\tfrac{1}{2}CO + 3\tfrac{1}{2}C + 2\tfrac{1}{2}H_2O + 1\tfrac{1}{2}N_2 \qquad (5.10)$$

One mole of TNT produces $7\tfrac{1}{2}$ moles of gas which will occupy 168 dm^3 and 1 g of TNT will produce 0.740 dm^3 g^{-1} (740 cm^3 g^{-1}) of gas at 'stp' as shown in Equation 5.11:

$7\tfrac{1}{2}$ mol of gas will occupy 22.4 × $7\tfrac{1}{2}$ = 168 dm^3

Therefore, 1 mol of TNT produces 201.6 dm^3 of gas

1 g of TNT produces $\dfrac{168}{227}$ = 0.740 dm^3 g^{-1}

where the molecular mass of TNT is 227 $\qquad (5.11)$

Table 5.13 presents the calculated volumes of gases produced by some explosive substances at 'stp'.

Table 5.13 *Calculated volume of gases produced by some explosive substances at standard temperature and pressure*

Explosive substance	Decomposition products	$V/\text{dm}^3\,\text{g}^{-1}$
Nitroglycerine	$3\,CO_2 + 2\frac{1}{2}\,H_2O + \frac{1}{2}\,O + 1\frac{1}{2}\,N_2$	0.740
EGDN	$2\,CO_2 + 2\,H_2O + N_2$	0.737
PETN	$2\,CO + 3\,CO_2 + 4\,H_2O + 2\,N_2$	0.780
RDX	$3\,CO + 3\,H_2O + 3\,N_2$	0.908
HMX	$4\,CO + 4\,H_2O + 4\,N_2$	0.908
Nitroguanidine	$CO + H_2O + H_2 + 2\,N_2$	1.077
Picric acid	$5\frac{1}{2}\,CO + \frac{1}{2}\,C + 1\frac{1}{2}\,H_2O + 1\frac{1}{2}\,N_2$	0.831
Tetryl	$5\frac{1}{2}\,CO + 1\frac{1}{2}\,C + 2\frac{1}{2}\,H_2O + 2\frac{1}{2}\,N_2$	0.820
TATB	$3\,CO + 3\,C + 3\,H_2O + 3\,N_2$	0.781
HNS	$9\,CO + 5\,C + 3\,H_2O + 3\,N_2$	0.747
TNT	$3\frac{1}{2}\,CO + 3\frac{1}{2}\,C + 2\frac{1}{2}\,H_2O + 1\frac{1}{2}\,N_2$	0.740

EXPLOSIVE POWER AND POWER INDEX

In an explosive reaction, heat and gases are liberated. The volume of gas V and the heat of explosion Q can both be calculated independently but these values can be combined to give the value for the explosive power as shown in Equation 5.12.

$$\text{Explosive Power} = Q \times V \qquad (5.12)$$

The value for the explosive power is then compared with the explosive power of a standard explosive (picric acid) resulting in the power index, as shown in Equation 5.13, where data for $Q_{(\text{picric acid})}$ and $V_{(\text{picric acid})}$ are 3250 kJ kg^{-1} and 0.831 dm^3 g^{-1}, respectively.

$$\text{Power Index} = \frac{Q \times V}{Q_{(\text{picric acid})} \times V_{(\text{picric acid})}} \times 100 \qquad (5.13)$$

The power index values of some primary and secondary explosive substances are presented in Table 5.14.

As expected, the values for the power and power index of secondary explosives are much higher than the values for primary explosives.

TEMPERATURE OF CHEMICAL EXPLOSION

When an explosive detonates the reaction is extremely fast and, initially, the gases do not have time to expand to any great extent. The heat

Table 5.14 *The power index of some primary and secondary explosive substances taking picric acid as the standard*

Explosive substance	$Q_v/\text{kJ g}^{-1}$	$V/\text{dm}^3\,\text{g}^{-1}$	$Q \times V \times 10^4$	*Power index/%*
Primary explosives				
Mercury fulminate	1755	0.215	37.7	14
Lead styphnate	1885	0.301	56.7	21
Lead azide	1610	0.218	35.1	13
Secondary explosives				
Nitroglycerine	6194	0.740	458.4	170
EGDN	6658	0.737	490.7	182
PETN	5794	0.780	451.9	167
RDX	5036	0.908	457.3	169
HMX	5010	0.908	454.9	169
Nitroguanidine	2471	1.077	266.1	99
Picric acid	3249	0.831	270.0	100
Tetryl	4335	0.820	355.5	132
TATB	3496	0.781	273.0	101
HNS	3942	0.747	294.5	109
TNT	4247	0.740	314.3	116

liberated by the explosion will raise the temperature of the gases, which will in turn cause them to expand and work on the surroundings to give a 'lift and heave effect'. The effect of this heat energy on the gas can be used to calculate the temperature of explosion.

The temperature of explosion T_e is the maximum temperature that the explosion products can attain under adiabatic conditions. It is assumed that the explosive at an initial temperature T_i is converted to gaseous products which are also at the initial temperature T_i. The temperature of these gaseous products is then raised to T_e by the heat of explosion Q. Therefore the value of T_e will depend on the value of Q and on the separate molar heat capacities of the gaseous products as shown in Equation 5.14, where C_v is the molar heat capacities of the products at constant volume and Σ represents the summation of the heat capacity integrals corresponding to the separate components of the gas mixture:

$$Q = \Sigma \int_{T_i}^{T_e} C_v \, dT \qquad (5.14)$$

The rise in temperature of the gases is calculated by dividing the heat generated Q by the mean molar heat capacity of the gases at constant volume ΣC_v as shown in Equation 5.15, where T_i is the initial temperature.

$$T_e = \frac{Q}{\Sigma C_v} + T_i \qquad (5.15)$$

Unfortunately, the heat capacities of the gaseous products vary with temperature in a non-linear manner and there is no simple relationship between temperature and C_v. The mean molar heat capacities of some gaseous products at various temperatures are presented in Table 5.15.

Using this information, the heat liberated by an explosion at various temperatures can be calculated. T_e can then be determined from a graphical representation of T_e versus Q, as shown in the example below.

The heat liberated Q during an explosion of RDX was calculated earlier to be 1118 kJ mol^{-1}. This heat is then used to raise the temperature of 1 mol of the gaseous products from their initial temperature T_i to their final temperature T_e. Let us assume that the temperature of the

Table 5.15 *Mean molar heat capacities at constant volume* J mol^{-1} K^{-1}

Temp/K	CO_2	CO	H_2O	H_2	N_2
2000	45.371	25.037	34.459	22.782	24.698
2100	45.744	25.204	34.945	22.966	24.866
2200	46.087	25.359	35.413	23.146	25.025
2300	46.409	25.506	35.865	23.322	25.175
2400	46.710	25.640	36.292	23.493	25.317
2500	46.991	25.769	36.706	23.665	25.451
2600	47.258	25.895	37.104	23.832	25.581
2700	47.509	26.012	37.485	23.995	25.703
2800	47.744	26.121	37.849	24.154	25.820
2900	47.965	26.221	38.200	24.309	25.928
3000	48.175	26.317	38.535	24.260	26.029
3100	48.375	26.409	38.861	24.606	26.129
3200	48.568	26.502	39.171	24.748	26.225
3300	48.748	26.589	39.472	24.886	26.317
3400	48.924	26.669	39.761	25.025	26.401
3500	49.091	26.744	40.037	25.158	26.481
3600	49.250	26.819	40.305	25.248	26.560
3700	49.401	26.891	40.560	25.405	26.635
3800	49.546	26.962	40.974	25.527	26.707
3900	49.690	27.029	41.045	25.644	26.778
4000	49.823	27.091	41.271	25.757	26.845
4500	50.430	27.372	42.300	26.296	27.154
5000	50.949	27.623	43.137	26.769	27.397

explosion T_e equals 3500 K and the initial temperature equals 300 K. The heat liberated by the explosion of RDX at a temperature of 3500 K is calculated from the mean molar heat capacities of the gaseous products at 3500 K as shown in Equation 5.16:

Mean molar heat capacities at 3500 K, using Table 5.15:

$\quad$ $CO = 26.744$ J mol^{-1}K^{-1}

$\quad$ $H_2O = 40.037$ J mol^{-1}K^{-1}

$\quad$ $N_2 = 26.481$ J mol^{-1} K^{-1}

Heat liberated by the explosion of
$\quad$ RDX at 3500 K $= Q_{3500\,K} = (\Sigma C_v) \times (T_e - T_i)$:

$\quad$ $Q_{3500\,K} = [(3 \times 26.744) + (3 \times 40.037) + (3 \times 26.481)] \times (3500 - 300)$

$\quad$ $Q_{3500\,K} = 895\,315$ J mol^{-1}

$\quad$ $Q_{3500\,K} = 895.3$ kJ mol^{-1} $\hfill$ (5.16)

The calculated value of $Q_{3500\,K}$ is 895.3 kJ mol^{-1} which is too low since the heat liberated during an explosion of RDX is already known to be 1118 kJ mol^{-1}. Therefore, the temperature of explosion must be higher. Let us now assume that T_e equals 4500 K, the heat liberated by the explosion of RDX at this higher temperature is calculated to be 1220.0 kJ mol^{-1} as shown in Equation 5.17:

Mean molar heat capacities at 4500 K:

$\quad$ $CO = 27.372$ J mol^{-1} K^{-1}

$\quad$ $H_2O = 42.300$ J mol^{-1} K^{-1}

$\quad$ $N_2 = 27.154$ J mol^{-1} K^{-1}

Heat liberated by explosion of RDX at 4500 K:

$\quad$ $Q_{4500\,K} = [(3 \times 27.372) + (3 \times 42.300) + (3 \times 27.154)] \times (4500 - 300)$

$\quad$ $Q_{4500\,K} = 1\,220\,007.6$ J mol^{-1}

$\quad$ $Q_{4500\,K} = 1220.0$ kJ mol^{-1} $\hfill$ (5.17)

This value of $Q_{4500\,K}$ is too high since the heat liberated during an explosion of RDX is 1118 kJ mol^{-1}. By calculating the values of Q at

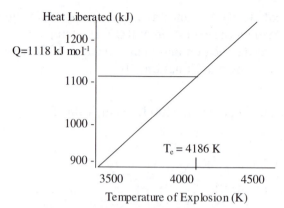

Figure 5.6 *Calculated values of heat liberated Q at various temperatures of explosion* T_e *for RDX*

various temperatures of T_e a graph can be drawn and the correct value for T_e can be obtained as shown in Figure 5.6.

Using the graph plotted in Figure 5.6 the value for the temperature of explosion for RDX is found to be $\sim$4186 K.

MIXED EXPLOSIVE COMPOSITIONS

Most explosive and propellant compositions contain a mixture of components so as to optimize their performance. Some of the components may not contribute to the heat liberated and may not even contain oxygen. These materials may however, contribute to the gaseous products and reduce the actual temperatures obtained on detonation of the explosive or burning of the propellant. An example of a typical mixed explosive composition is one which contains 60% RDX and 40% TNT, and where the heat of explosion Q has been optimized. In order to calculate the values of Q and V for this composition the oxygen balance and the reaction for decomposition need to be determined. But even before these can be calculated the atomic composition of the mixture must first be established.

Atomic Composition of the Explosive Mixture

Calculations for the atomic composition of 60% RDX/40% TNT are presented in Figure 5.7.

From the calculations of the atomic composition the empirical formula is found to be $C_{0.0204}H_{0.0250}N_{0.0215}O_{0.0268}$.

Explosive substance	Empirical formula	Proportion by mass (g)	Molar mass (g)	Molar proportion in 1g of mixture	Mol of atoms in 1g of explosive mixture			
					C	H	N	O
RDX	$C_3H_6N_6O_6$	0.6	222	$\dfrac{0.6}{222} = 0.00270$	0.00810	0.01620	0.01620	0.01620
TNT	$C_7H_5N_3O_6$	0.4	227	$\dfrac{0.4}{227} = 0.00176$	0.01232	0.00880	0.00528	0.01056
				Total	0.02042	0.02500	0.02148	0.02676

Figure 5.7 *Calculations for the atomic composition of 60% RDX/40% TNT*

Oxygen Balance

The oxygen balance can now be calculated as shown in Equation 5.18:

$$\Omega = \frac{[d - (2a) - (b/2)] \times 1600}{M}$$

$$\Omega = \frac{[0.0268 - (2 \times 0.0204) - (0.0250/2)] \times 1600}{1} = -42 \quad (5.18)$$

For binary mixtures the calculated value of Ω is directly proportional to the amount of each component and their individual Ω values as shown in Equation 5.19,

$$\Omega = \Omega_1 f_1 + (1 - f_1)\Omega_2$$
$$\Omega = (-21.6 \times 0.6) + (1 - 0.6)(-74)$$
$$\Omega = -42 \quad (5.19)$$

where Ω_1 and f_1 are the oxygen balance and mass of the first component (*i.e.* RDX), respectively, and Ω_2 is the oxygen balance of the second component, (*i.e.* TNT).

Decomposition Reaction

The reaction for the decomposition products of the mixed explosive composition can be determined by applying the modified Kistiakowsky–Wilson rules as shown in Table 5.16.

The overall reaction for the decomposition of 60% RDX/40% TNT is given in Reaction 5.11.

$$C_{0.0204}H_{0.0250}N_{0.0215}O_{0.0268} \rightarrow 0.0143CO + 0.0061C + 0.0125H_2O + 0.01075N_2$$
$$(5.11)$$

Table 5.16 *Decomposition products of 60% RDX/40% TNT using the Kistiakowsky–Wilson rules*

Rule no.	Conditions	Products
1	Hydrogen atoms are converted to water	$0.0250\,H \rightarrow 0.0125\,H_2O$
2	If any oxygen remains then carbon is converted to carbon monoxide	$0.0143\,O \rightarrow 0.0143\,CO$
3	If any oxygen still remains then carbon monoxide is oxidized to carbon dioxide	No more oxygen
4	All the nitrogen is converted to nitrogen gas, N_2	$0.0215\,N \rightarrow 0.01075\,N_2$

Heat of Explosion

The heat of explosion for the mixed explosive composition can now be calculated from the heat of formation for the explosive components and the heat of formation for its products as shown in Figure 5.8.

Using the diagram in Figure 5.8 the heat of detonation for 60% RDX/40% TNT can be calculated as shown in Equation 5.20:

$$\Delta H_d = \Delta H_2 - \Delta H_1$$

$$\Delta H_f(\text{RDX}) = +62.0 \text{ kJ mol}^{-1}$$

$$\Delta H_f(\text{TNT}) = -26 \text{ kJ mol}^{-1}$$

$$\Delta H_f(\text{CO}) = -110.0 \text{ kJ mol}^{-1}$$

$$\Delta H_f(\text{H}_2\text{O}_{(g)}) = -242.0 \text{ kJ mol}^{-1}$$

$$\Delta H_1 = \Delta H_f(\text{C}_{0.0204}\text{H}_{0.0250}\text{N}_{0.0215}\text{O}_{0.0268})$$

$$\Delta H_1 = [0.0027 \times \Delta H_f(\text{RDX})] + [0.00176 \times \Delta H_f(\text{TNT})]$$

$$\Delta H_1 = [0.0027 \times (+62)] + [0.00176 \times (-26)] = +0.122 \text{ kJ}$$

$$\Delta H_2 = [0.0143 \times \Delta H_f(\text{CO})] + [0.0125 \times \Delta H_f(\text{H}_2\text{O}_{(g)})]$$

$$\Delta H_2 = [0.0143 \times (-110.0)] + [0.0125 \times (-242.0)] = -4.598 \text{ kJ}$$

$$\Delta H_d = \Delta H_2 - \Delta H_1 = -4.598 - (+0.122) = -4.72 \text{ kJ} \qquad (5.20)$$

The calculated value for the heat of detonation for 1 g of 60% RDX/40% TNT is -4.72 kJ. The heat of explosion Q is calculated to be 4720 kJ kg^{-1} as shown in Equation 5.21 where M is 1 g:

$$Q = \frac{\Delta H_d \times 1000}{M} = \frac{-4.72 \times 1000}{1} = -4720 \text{ kJ kg}^{-1} \qquad (5.21)$$

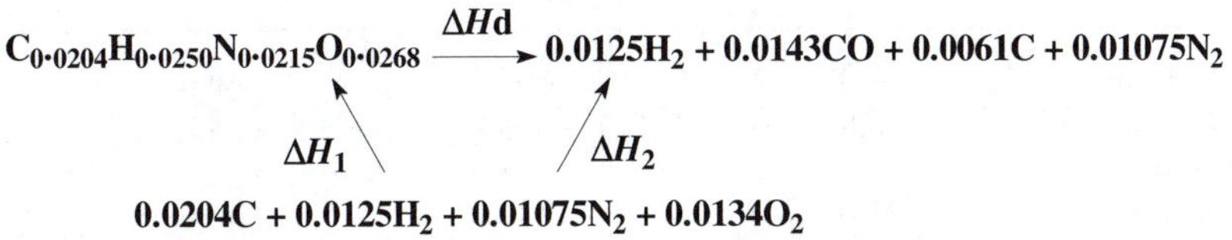

$$\text{C}_{0.0204}\text{H}_{0.0250}\text{N}_{0.0215}\text{O}_{0.0268} \xrightarrow{\Delta Hd} 0.0125\text{H}_2 + 0.0143\text{CO} + 0.0061\text{C} + 0.01075\text{N}_2$$

$$\underset{\Delta H_1}{\searrow} \qquad \underset{\Delta H_2}{\nearrow}$$

$$0.0204\text{C} + 0.0125\text{H}_2 + 0.01075\text{N}_2 + 0.0134\text{O}_2$$

Figure 5.8 *Enthalpy of detonation for an explosive mixture containing 60% RDX and 40% TNT*

Volume of Gaseous Products

The volume of the gaseous products liberated during an explosion by the mixed explosive composition at 'stp' can be calculated from the equation for explosion as shown in Reaction 5.12.

$$C_{0.0204}H_{0.0250}N_{0.0215}O_{0.0268} \rightarrow 0.0143CO + 0.0061C + 0.0125H_2O + 0.01075N_2$$
(5.12)

From Reaction 5.12 it can be seen that 0.03755 mol of gas is produced from 1 g of mixed explosive composition, which will occupy a volume of $0.841 \text{ dm}^3 \text{ g}^{-1}$ as shown in Equation 5.22:

1 g of gaseous product contains 0.03755 mol

1 g of gaseous product will occupy $22.4 \times 0.03755 = 0.841 \text{ dm}^3 \text{ g}^{-1}$
(5.22)

The mixed explosive composition containing 60% RDX with 40% TNT therefore has a Q value of 4720 kJ kg^{-1}, a V value of $0.841 \text{ dm}^3 \text{ g}^{-1}$ and a power index equal to 147%.

ENERGIZED EXPLOSIVES

The heat of explosion Q can be increased by adding to the explosive composition another fuel which has a high heat of combustion ΔH_c. Such fuels can be found with the lighter elements of the periodic table as shown in Table 5.17.

Beryllium has the highest heat of combustion of the solid elements, followed by boron and aluminium. Aluminium is a relatively cheap and useful element, and is used to increase the performance of explosive compositions, such as aluminized ammonium nitrate and aluminized

Table 5.17 *Heats of combustion of some light elements*

Element	Relative atomic mass	ΔH_c /kJ mol^{-1}	ΔH_c /kJ g^{-1}
Beryllium	9.0	−602.1	−66.9
Boron	10.8	−635.0	−58.8
Lithium	6.9	−293.9	−42.6
Aluminium	27.0	−834.3	−30.9
Magnesium	24.3	−602.6	−24.8
Sulfur	32.1	−295.3	−9.2
Zinc	65.4	−353.2	−5.4

TNT. Aluminium is also used in some commercial blasting explosives, particularly in water-based slurry explosives, which contain a high percentage of ammonium nitrate.

Addition of Aluminium

The oxidation of aluminium is highly exothermic producing -1590 kJ, as shown in Reaction 5.13.

$$2Al_{(s)} + 1\frac{1}{2}O_{2(g)} \rightarrow Al_2O_{3(s)} \qquad \Delta H_c = -1590 \text{ kJ} \qquad (5.13)$$

In an explosive composition the aluminium reacts with the gaseous products particularly in oxygen-deficient compositions where no free oxygen exists as shown in Reaction 5.14.

$$3CO_{2(g)} + 2Al_{(s)} \rightarrow 3CO_{(g)} + Al_2O_{3(s)} \qquad \Delta H_c = -741 \text{ kJ}$$

$$3H_2O_{(g)} + 2Al_{(s)} \rightarrow 3H_{2(g)} + Al_2O_{3(s)} \qquad \Delta H_c = -866 \text{ kJ}$$

$$3CO_{(g)} + 2Al_{(s)} \rightarrow 3C_{(s)} + Al_2O_{3(s)} \qquad \Delta H_c = -1251 \text{ kJ} \quad (5.14)$$

The volume of gas does not change in the first two reactions, *i.e.* 3 moles → 3 moles. Consequently, the increase in the output of heat from the oxidation of aluminium prolongs the presence of high pressures. This effect is utilized in explosive compositions for airblasts, lifting and heaving, or large underwater bubbles. However, there is a limit to the amount of aluminium that can be added to an explosive composition as shown in Table 5.18.

The heat of explosion Q increases with an increase in the quantity of aluminium but the gas volume V decreases, resulting in the power $Q \times V$ reaching a maximum value of 381×10^{-4} at 18% aluminium.

Table 5.18 *Effect of the addition of aluminium on the heat of explosion and volume of gaseous products for TNT/Al. These values have been obtained experimentally*

Aluminium/% weight	Q/kJ kg^{-1}	V/dm^3 g^{-1}	$Q \times V \times 10^4$
0	4226	0.750	318
9	5188	0.693	360
18	6485	0.586	381
25	7280	0.474	343
32	7657	0.375	289
40	8452	0.261	222

The same effect can be observed for other explosive compositions containing aluminium, where a maximum value for the power can be achieved by adding 18–25% aluminium.

FORCE AND PRESSURE OF EXPLOSION

This chapter has so far described the total chemical energy released when a chemical explosion takes place. This energy is released in the form of kinetic energy and heat over a very short time, *i.e.* microseconds. In a detonating explosive a supersonic wave is established near to the initiation point and travels through the medium of the explosive, sustained by the exothermic decomposition of the explosive material behind it. On reaching the periphery of the explosive material the detonation wave passes into the surrounding medium, and exerts on it a sudden, intense pressure, equivalent to a violent mechanical blow. If the medium is a solid, *i.e.* rock or stone, the violent mechanical blow will cause multiple cracks to form in the rock. This effect is known as 'brisance' which is directly related to the detonation pressure in the shockwave front.

After the shockwave has moved away from the explosive composition the gaseous products begin to expand and act upon the surrounding medium. A crater will be formed if the medium is earth, in water a gas bubble is formed and in air a blast wave develops. The intensity of the gaseous expansion will depend upon the power ($Q \times V$) of the explosive.

The pressure of explosion P_e is the maximum static pressure which may be achieved when a given weight of explosive is burned in a closed vessel of fixed volume. The pressure attained is so high that the Ideal Gas Laws are not sufficiently accurate and have to be modified by using a co-volume α. At high pressure

$$P_e(V_i - \alpha) = nRT_e \tag{5.23}$$

where V_i is the volume of the closed test vessel, n is the number of moles of gas produced per gram of explosive, T_e is the temperature of explosion in Kelvin and R is the molar gas constant. For example, the maximum pressure of an explosion P_e inside a vessel can be calculated when 10 g of RDX is detonated in a vessel of volume 10 cm^3 as shown in Equation 5.24. Here the number of moles of gas n produced per gram of RDX is 0.0405, the temperature of explosion is 4255K and the co-volume α equals 0.55.

$$P_e = \frac{(0.0405 \times 8.314 \times 4255)}{(1 - 0.55 \times 10^{-6}}$$

$$P_e = 3183.85 \times 10^6 \text{ N m}^{-2}$$

$$P_e = 3183.85 \text{ MN m}^{-2} \tag{5.24}$$

The pressure of explosions are of a much lower order of magnitude than the detonation pressures.

The power of high explosives can be expressed in terms of their 'force' and has the units of kJ g^{-1} or MJ kg^{-1}, which is the same units as the power of explosion. The 'force' can be calculated from Equation 5.25, where F is the force constant, n is the number of moles of gas produced per gram of explosive, R is the molar gas constant and T_e is the temperature of explosion in Kelvin.

$$F = nRT_e \tag{5.25}$$

The force exerted by the gases for the detonation of RDX is calculated in Equation 5.26, where number of moles of gas n produced per gram of RDX is 0.0405 and the temperature of explosion is 4255K.

$$F = 0.0405 \times 8.314 \times 4255$$

$$F = 1432.7 \text{ J g}^{-1} \tag{5.26}$$

The values of F for high explosives tend to be higher than those for gun propellants, because the latter are designed to have lower combustion temperatures. This is not so with the more energetic types of rocket propellants where the F values may exceed those of high explosives. In general, explosion products of low molar mass favour high values of n and therefore have high values of F.

The 'force' and 'power of explosion' are in fact the same, as shown below in Equation 5.27 and Equation 5.28.

$$F = nRT_e$$

$$\text{If } n = \frac{\text{volume of gaseous products at stp}}{\text{molar volume of gas at stp}} = \frac{V}{22.4}$$

$$\text{and } T_e = \left(\frac{Q}{\Sigma C_v} + T_i \right)$$

$$\text{then } F = \frac{V}{22.4} R \left(\frac{Q}{\Sigma C_v} + T_i \right) \tag{5.27}$$

where Q is the heat of explosion, ΣC_v is the sum of the mean molar heat capacities and T_i is the initial temperature. Let k replace the values of the molar volume of a gas at stp and the molar gas constant, and assume that ΣC_v is the sum of the mean molar heat capacities of the products between 300K and T_e. Then

$$F = k\left(\frac{QV}{\Sigma C_v}\right)$$
(5.28)

The term $(Q_v/\Sigma C_v)$ is called the 'characteristics product' of the explosive. For most explosives, ΣC_v may be taken as constant as shown in equation 5.29.

$$F = kQV$$

$$F \propto QV$$
(5.29)

Chapter 6

Equilibria and Kinetics of Explosive Reactions

The thermochemistry of explosive compositions has been discussed in detail in Chapter 5. The Kistiakowsky–Wilson and the Springall Roberts rules both give an approximate estimate for the products of decomposition, which is independent of the temperature of explosion. The formulae and calculations for determining the heat of explosion also assume that the explosive reactions go to total completion. However, in practice the reactions do not go to completion and an equilibrium is set up between the reactants and the products. This equilibrium is also dependent upon the temperature of the explosion T_e.

EQUILIBRIA

During a chemical explosive reaction many equilibria take place. These equilibria are dependent on the oxygen balance of the system. The most important equilibria are presented in Reaction 6.1.

$$CO_2 + H_2 \rightleftharpoons CO + H_2O$$

$$2CO \rightleftharpoons C + CO_2$$

$$CO + H_2 \rightleftharpoons C + H_2O \qquad (6.1)$$

The last equation in Reaction 6.1 only becomes important as the oxygen balance decreases and the quantity of carbon monoxide and hydrogen in the products begins to increase. All three equilibria are temperature-dependent, and the equilibrium position will move to favour the reactants or products depending on the temperature generated by the reaction.

103

Products of Decomposition

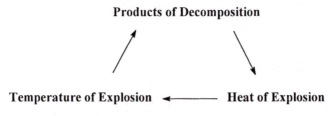

Temperature of Explosion ←——— **Heat of Explosion**

Figure 6.1 *Cycle of determinable parameters for equilibrium reactions*

In order to determine the products of decomposition for an equilibrium reaction the temperature of explosion T_e is required. T_e can be calculated from the heat of explosion Q which in turn depends upon the products of decomposition as shown in Figure 6.1.

In order to determine the products of decomposition for equilibrium reactions the Kistiakowsky–Wilson or the Springall Roberts rules can be applied as a starting point. From the products of decomposition the heat and temperature of explosion can then be calculated. The temperature of explosion can then be used to calculate the products of decomposition. In practice, this process is repeated many times until there is agreement between the answers obtained. Equilibria of complex reactions and of multi-component systems are today obtained by computer; however, the ability to use tabulated data is useful in predicting the direction and extent of the reaction.

The example below will evaluate the products of decomposition, temperature of explosion and heat of explosion for RDX at equilibrium conditions.

Products of Decomposition

As a starting point the decomposition products for RDX $(C_3H_6N_6O_6)$ are calculated from the Kistiakowsky–Wilson rules $(\Omega = -21.60)$ as shown in Reaction 6.2.

$$C_3H_6N_6O_6 \rightarrow 3CO + 3H_2O + 3N_2 \tag{6.2}$$

The complete decomposition of RDX does not actually take place because there is not sufficient oxygen in the RDX molecule for complete oxidation of the reactants. Instead, the products compete for the available oxygen and various equilibrium reactions are set up. The most important of these equilibrium reactions is the water–gas equilibrium as shown in Reaction 6.3.

$$CO_2 + H_2 \rightleftharpoons CO + H_2O \tag{6.3}$$

The Water–Gas Equilibrium

The water–gas equilibrium is present not only in the decomposition of RDX but in many other chemical explosive reactions. The temperature of the reaction will move the equilibrium to the left or right. An increase in the temperature of the reaction will result in hydrogen becoming more successful in competing for the oxygen, whereas at low temperatures carbon is more successful.

The total amount of energy liberated, *i.e.* the heat of explosion Q in the water–gas equilibrium, depends upon the relative proportions of the reactants (carbon dioxide and hydrogen) to the products (carbon monoxide and water). This can be seen from Equation 6.1 where the heat of formation for carbon dioxide emits more energy than that for carbon monoxide and water:

$$\Delta H_f(CO_2) = -393.7 \text{ kJ mol}^{-1}$$

$$\Delta H_f(CO) + \Delta H_f(H_2O_{(g)}) = -110.0 + -242.0 = -352.0 \text{ kJ mol}^{-1} \quad (6.1)$$

On the other hand, the heat capacities of carbon monoxide and water are less than those for carbon dioxide and hydrogen as shown in Table 6.1.

Consequently, the small amount of heat generated by the formation of carbon monoxide and water may result in a large amount of heat of explosion.

Heat of Explosion

In order to calculate the heat of explosion for RDX at equilibrium conditions the exact quantities of carbon dioxide, carbon monoxide and water must be determined. These can be calculated from the equilibrium

Table 6.1 *Comparison of the mean molar heat capacities for carbon dioxide and hydrogen, and carbon monoxide and water*

Temp/K	$CO_2 + H_2$/J mol^{-1} K^{-1}	$H_2O + CO$/J mol^{-1} K^{-1}
2000	68.153	59.496
2500	70.656	62.475
3000	72.435	64.852
3500	74.249	66.781
4000	75.580	68.362
4500	76.726	69.672
5000	77.718	70.760

constant. The equilibrium constant K_1 for the water–gas equilibrium is given in Equation 6.2, where the terms in the square brackets [ ] are the concentrations in mol dm^{-3}.

$$K_1 = \frac{[CO]\,[H_2O]}{[CO_2]\,[H_2]} \tag{6.2}$$

By using the general formula of an explosive $C_aH_bN_cO_d$ the concentrations for the reactants and products can be determined. For the reaction involving RDX, carbon will react to form carbon monoxide and carbon dioxide, hydrogen will react to form hydrogen gas and water, nitrogen will react to form nitrogen gas, and oxygen will combine with other elements in the explosive composition to form carbon monoxide, carbon dioxide and water as shown in Reaction 6.4.

$$C_aH_bN_cO_d \rightarrow n_1CO_2 + n_2H_2O + n_3N_2 + n_4CO + n_5H_2 \tag{6.4}$$

The values for the terms a, b, c and d in the reactant can be determined from the stoichiometric equations involving the number of moles, *i.e.* n_1, n_2, n_3, *etc.*, for the products. The equilibrium equation in terms of n_1, n_2, n_4 and n_5 and the stoichiometric equations are presented in Equations 6.3 and 6.4, respectively.

$$K_1 = \frac{n_4\,n_2}{n_1\,n_5} \tag{6.3}$$

$a = n_1 + n_4$ (carbon-containing molecules)

$b = 2n_2 + 2n_5$ (hydrogen-containing molecules)

$c = 2n_3$ (nitrogen-containing molecules)

$d = 2n_1 + n_2 + n_4$ (oxygen-containing molecules) (6.4)

The formulae shown in Equation 6.4 can be rearranged in terms of n_1, n_2, n_4 and n_5 as shown in Equation 6.5.

$n_4 = a - n_1$

$n_2 = d - 2n_1 - n_4$

Substituting for n_4 gives $n_2 = d - a - n_1$

$n_5 = b/2 - n_2$

Elimination of n_2 gives $n_5 = b/2 - d + a + n_1$ (6.5)

The equation for the equilibrium can be now be written in terms of n_1, a, b, and d as shown in Equation 6.6.

$$K_1 = \frac{(a - n_1)(d - a - n_1)}{(n_1)(b/2 - d + a + n_1)} \qquad (6.6)$$

The value of K_1 increases with temperature as shown in Table 6.2.

Therefore, the products on the right-hand side of the water–gas equilibrium will increase as the temperature rises. If we assume that the temperature of explosion $T_e = 4000$ K, then the value of K_1 becomes 9.208. By substituting the values for K_1, a, b and d into Equation 6.6 the concentration of CO_2 can be calculated as shown in Equation 6.7, where

Table 6.2 *The dependence of the equilibrium constant on temperature for the water–gas reaction*

Temperature/K	Equilibrium constant K_1
1600	3.154
1700	3.584
1800	4.004
1900	4.411
2000	4.803
2100	5.177
2200	5.533
2300	5.870
2400	6.188
2500	6.488
2600	6.769
2700	7.033
2800	7.279
2900	7.509
3000	7.723
3100	7.923
3200	8.112
3300	8.290
3400	8.455
3500	8.606
3600	8.744
3700	8.872
3800	8.994
3900	9.107
4000	9.208
4500	9.632
5000	9.902

the values of a, b and d for RDX ($C_3H_6N_6O_6$) are 3, 6 and 6, respectively.

$$9.208 = \frac{(3 - n_1)(6 - 3 - n_1]}{(n_1)(6/2 - 6 + 3 + n_1)}$$

$$9.208 = \frac{18 - 9 - 3n_1 - 6n_1 + 3n_1 + n_1{}^2}{n_1{}^2}$$

$$9.208 = \frac{9 - 6n_1 + n_1{}^2}{n_1{}^2}$$

$$9.208n_1{}^2 - 9 + 6n_1 - n_1{}^2 = 0$$

$$8.208n_1{}^2 + 6n_1 - 9 = 0$$

$$n_1 = 0.7436 \text{ or } -1.4764 \tag{6.7}$$

The quadratic equation results in two values for the concentration of n_1, one of which is negative and can be ignored. The concentrations of the other compounds can now be determined by substituting the 0.7436 for n_1 as shown in Equation 6.8.

$$n_4 = 3 - 0.7436 = 2.2564 = [CO]$$

$$n_2 = 6 - 3 - 0.7436 = 2.2564 = [H_2O]$$

$$n_5 = 3 - 6 + 3 + 0.7436 = 0.7436 = [H_2] \tag{6.8}$$

Using the values calculated in Equation 6.8 the value of K_1 can be obtained, as shown in Equation 6.9.

$$K_1 = \frac{2.2564 \times 2.2564}{0.7436 \times 0.7436} = 9.20775 \tag{6.9}$$

This value compares well with the value of 9.208 used in the initial calculations.

The overall equation for the detonation of RDX at 4000 K under equilibrium conditions is given in Reaction 6.5.

$$C_3H_6N_6O_6 \rightarrow 0.7436CO_2 + 2.2564H_2O + 2.2564CO + 0.7436H_2 + 3N_2 \tag{6.5}$$

In order to calculate the heat of explosion Q for RDX the heat of detonation must first be determined using Hess's Law as shown in Figure 6.2.

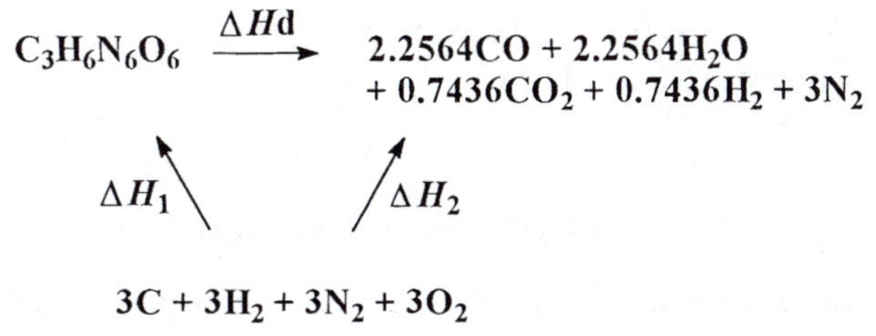

$$C_3H_6N_6O_6 \xrightarrow{\Delta Hd} 2.2564CO + 2.2564H_2O + 0.7436CO_2 + 0.7436H_2 + 3N_2$$

$$\Delta H_1 \nwarrow \qquad \nearrow \Delta H_2$$

$$3C + 3H_2 + 3N_2 + 3O_2$$

Figure 6.2 *Enthalpy of detonation for RDX under equilibrium conditions using Hess's law*

Using the diagram in Figure 6.2 the heat of detonation for RDX under equilibrium conditions can be calculated as shown in Equation 6.10.

$$\Delta H_2 = \Delta H_1 + \Delta H_d \Rightarrow \Delta H_d = \Delta H_2 - \Delta H_1$$

$$\Delta H_f(RDX) = +62.0 \text{ kJ mol}^{-1}$$

$$\Delta H_f(CO_2) = -393.7 \text{ kJ mol}^{-1}$$

$$\Delta H_f(CO) = -110.0 \text{ kJ mol}^{-1}$$

$$\Delta H_f(H_2O_{(g)}) = -242.0 \text{ kJ mol}^{-1}$$

Therefore, $\Delta H_1 = \Delta H_f(RDX) = +62.0 \text{ kJ mol}^{-1}$

$$\Delta H_2 = [2.2564 \times \Delta H_f(CO)] + [2.2564 \times \Delta H_f(H_2O_{(g)})] + [0.7436 \times \Delta H_f(CO_2)]$$

$$\Delta H_2 = [2.2564 \times (-110.0)] + [2.2564 \times (-242.0)] + [0.7436 \times (-393.7)]$$

$$\Delta H_2 = -1087.0 \text{ kJ mol}^{-1}$$

$$\Delta H_d = \Delta H_2 - \Delta H_1 = -1087.0 - (+62.0) = -1149.0 \text{ kJ mol}^{-1}$$

$$(6.10)$$

The heat of detonation for RDX is $-1149 \text{ kJ mol}^{-1}$. This value can be converted to the heat of explosion Q as shown in Equation 6.11, where M is the molar mass of RDX.

$$Q = \frac{\Delta H_d \times 1000}{M} = \frac{-1149 \times 1000}{222} = -5176 \text{ kJ kg}^{-1} \quad (6.11)$$

The heat of explosion for RDX at equilibrium conditions is found to be 5176 kJ kg^{-1}; this is higher than the value calculated in Chapter 5 using

the Kistiakowsky–Wilson approach (5036 kJ kg^{-1}) which assumes that
the reaction goes to completion.

Temperature of Explosion

In calculating the heat of explosion we assumed that the temperature of
explosion T_e was 4000 K. If this assumption was correct then the heat
liberated by the explosion at 4000 K should equal 1149 kJ mol^{-1}. The
calculated value for the heat liberated, where the initial temperature is
taken as 300 K, is presented in Equation 6.12. The values for the mean
molar heat capacities at constant volume can be found in Table 5.15.

Mean molar heat capacities at 4000 K:

$CO = 27.091$ J mol^{-1} K^{-1}

$H_2O = 41.271$ J mol^{-1} K^{-1}

$CO_2 = 49.823$ J mol^{-1} K^{-1}

$H_2 = 25.757$ J mol^{-1} K^{-1}

$N_2 = 26.845$ J mol^{-1} K^{-1}

Heat liberated by the explosion of RDX at
    4000 K $= Q_{4000\,K} = (\Sigma C_v) \times (T_e - T_i)$:

$Q_{4000\,K} = [(2.2564 \times 27.091) + (2.2564 \times 41.271) + (0.7436 \times 49.823)$
$+ (0.7436 \times 25.757) + (3 \times 26.845)] \times (4000 - 300)$

$Q_{4000\,K} = 1\,076\,656.7$ J mol^{-1}

$Q_{4000\,K} = 1077$ kJ mol^{-1} (6.12)

The calculated heat liberated at 4000 K is 1077 kJ mol^{-1}, which is lower
than the original value of 1149 kJ mol^{-1}. A higher temperature of
explosion should now be taken and the cycle repeated. Eventually this
would lead to an answer of 4255 K for the temperature of explosion.
    Similar calculations can be carried out on mixed explosives and on
propellants if the composite formula is known for the materials. The use
of the water–gas equilibrium will produce a result that is a closer
approximation to the experimentally-derived figure than the Kistiakowsky–Wilson and the Springall Roberts approaches.
    The method used above for calculating the temperature and heat of
explosion for the water–gas equilibrium can also be applied to other

equilibrium equations. Examples of such equilibria which may become important during an explosive reaction are presented in Reaction 6.6.

$$\frac{1}{2}N_2 + CO_2 \rightleftharpoons CO + NO$$

$$2CO \rightleftharpoons C + CO_2 \qquad\qquad (6.6)$$

KINETICS OF EXPLOSIVE REACTIONS

Kinetics is the study of the rate of change of chemical reactions. These reactions can be very fast, *i.e.* instantaneous reactions such as detonation, those requiring a few minutes, *i.e.* dissolving sugar in water, and those requiring several weeks, *i.e.* the rusting of iron. In explosive reactions the rate is very fast and is dependent on the temperature and pressure of the reaction, and on the concentration of the reactants.

Activation Energy

During an explosive reaction, energy is first supplied from the initiator to raise the temperature of the explosive so that ignition takes place, with the formation of hotspots. If the energy generated by the hotspots is less than the activation energy no reaction will take place and the hotspots will gradually die out, as shown in Figure 6.3.

On the other hand, if the energy generated by the hotspots is greater

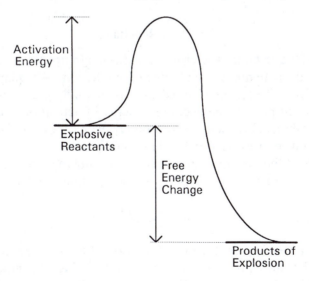

Figure 6.3 *Schematic diagram of the reaction profile for chemical explosives*

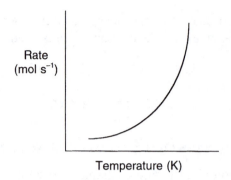

Figure 6.4 *The effect of temperature on the rate of reaction*

than the activation energy the reaction will proceed forward with the formation of the explosive products and the release of energy. The activation energy therefore represents an amount of energy which is required to take a starting material (*i.e.* an explosive compound at ambient temperature) and convert it to a reactive, higher-energy excited state. In this excited state, a reaction occurs to form the products with the liberation of a considerable amount of energy which is greater than the energy of activation.

The values for the activation energy can be used to measure the ease with which an explosive composition will initiate, where the larger the activation energy the more difficult it will be to initiate the explosive composition.

Rate of Reaction

The rate of the reaction is determined by the magnitude of the activation energy, and the temperature at which the reaction takes place. As the temperature of the system is raised, an exponentially-greater number of molecules will possess the necessary energy of activation. The rate of reaction will therefore increase accordingly in an exponential fashion as the temperature rises, as illustrated in Figure 6.4.

The rate of the reaction can be described using the rate–temperature relationship, which is known as the Arrhenius equation, 6.13,

$$k = Ae^{-E/RT} \tag{6.13}$$

where k is a constant for the rate of reaction, A is a constant for a given material, E is the activation energy in kJ mol^{-1}, T is the temperature in Kelvin and R is the Universal Gas constant, *i.e.* 8.314 J mol^{-1} K^{-1}. The constant A is known as the frequency factor or pre-exponential factor

Table 6.3 *Values for the activation energy* E *and collision factor* A *for some primary and secondary explosive substances*

Explosive substance	Activation energy $E/\text{kJ mol}^{-1}$	Collision factor A
Primary explosives		
Mercury fulminate	105	10^{11}
Silver azide	167	
Lead azide	160	
Secondary explosives		
Nitroglycerine	176	10^{19}
Tetryl (crystalline)	217	$10^{22.5}$
PETN	196	$10^{19.8}$
RDX	199	$10^{18.5}$
Picric acid	242	$10^{22.5}$
TNT	222	10^{19}
HMX	220	$10^{19.7}$

and can be determined from the number of collisions per unit volume per second between the molecules. The term $e^{-E/RT}$ is a measure of the fraction of colliding molecules that result in a reaction. Therefore, if $E = 0$ (*i.e.* zero activation energy) then $e^{-E/RT} = 1$ and all colliding molecules react; conversely, if $E/RT \gg 1$ then $E \gg RT$ (*i.e.* high activation energy), the majority of the colliding molecules do not react. The values for the collision factor A and the activation energy E for some primary and secondary explosives are presented in Table 6.3.

Primary explosives have low values for the activation energy and collision factor compared with secondary explosives. Therefore, it takes less energy to initiate primary explosives and makes them more sensitive to an external stimulus, *i.e.* impact, friction, *etc.*, whereas secondary explosives have higher values for the activation energy and collision factor, and are therefore more difficult to initiate and less sensitive to external stimulus.

Kinetics of Thermal Decomposition

All explosives undergo thermal decomposition at temperatures far below those at which explosions take place. These reactions are important in determining the stability and shelf life of the explosive. The reactions also provide useful information on the susceptibility of explosives to heat. The kinetic data are normally determined under isothermal conditions by measuring the rate of gas released from a sample of explosive at a series of controlled temperatures.

Equation 6.14 shows the relationship between the rate of decomposition and temperature, where V is the volume of gas evolved, T is the temperature in $°C$, k is the reaction rate constant, and C is a constant for the particular explosive.

$$V = k^T + C \tag{6.14}$$

For every $10\,°C$ increase in temperature, the rate of decomposition is approximately doubled, but may increase as much as 50 times if the explosive is in the molten state. The rates of decomposition depend on the condition of storage and the presence of impurities which may act as catalysts. For example, nitroglycerine and nitrocellulose decompose at an accelerated rate due to autocatalysis, whereas the decomposition rate of TNT, picric acid and tetryl can be reduced by removing the impurities which are usually less stable than the explosive itself. With many of the explosives the presence of moisture increases the rate of decomposition.

MEASUREMENT OF KINETIC PARAMETERS

The common methods of investigating the kinetics of explosive reactions are differential thermal analysis, thermogravimetric analysis and differential scanning calorimetry.

Differential Thermal Analysis

Differential thermal analysis (DTA) involves heating (or cooling) a test sample and an inert reference sample under identical conditions and recording any temperature difference which develops between them. Any physical or chemical change occurring to the test sample which involves the evolution of heat will cause its temperature to rise temporarily above that of the reference sample, thus giving rise to an exothermic peak on a DTA plot. Conversely, a process which is accompanied by the absorption of heat will cause the temperature of the test sample to lag behind that of the reference material, leading to an endothermic peak.

The degree of purity of an explosive can be determined from DTA plots. Contamination of the explosive will cause a reduction in the melting point. Consequently, the magnitude of the depression will reflect the degree of contamination. A phase change or reaction will give

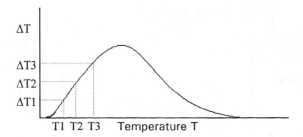

Figure 6.5 *DTA thermogram for the decomposition of an explosive material*

rise to an endothermic or exothermic peak, and the area under the peak is related to the amount of heat evolved or taken in.

Figure 6.5 shows the DTA plot for the decomposition of an explosive material. Decomposition begins when the temperature T reaches the ignition temperature of the explosive material resulting in an exothermic peak. As the explosive material decomposes it releases heat which is measured on the DTA plot as ΔT (y-axis). ΔT is proportional to the rate at which the explosive material decomposes; as the rate increases more heat is emitted and ΔT increases. Assuming that the decomposition of an explosive is a first order process, then the rate of the reaction is directly proportional to the increase in temperature ΔT, as shown in Equation 6.15.

$$\text{Rate of reaction} \propto \Delta T \qquad (6.15)$$

The activation energy E can be calculated from the DTA plot using the Arrhenius equation as shown in Equation 6.16.

$$\text{Rate of reaction} = \Delta T = Ae^{-E/RT}$$

$$\ln \Delta T = \ln (Ae^{-E/RT}) = \ln A + \ln e^{-E/RT}$$

$$\ln \Delta T = \ln A - E/RT \qquad (6.16)$$

ΔT is measured at several temperatures T, and $\ln \Delta T$ versus $1/T$ is plotted. A straight line is obtained with a slope of $-E/R$.

DTA will also provide information on the melting, boiling, crystalline transitions, dehydration, decomposition, oxidation and reduction reactions.

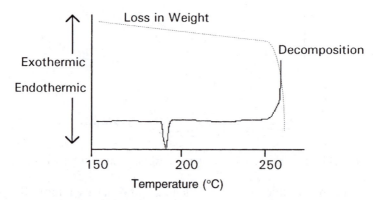

Figure 6.6 *Thermogram of HMX-β using differential thermal analysis and thermogravimetric analysis*

Thermogravimetric Analysis

Thermogravimetric analysis (TGA) is a suitable technique for the study of explosive reactions. In TGA the sample is placed on a balance inside an oven and heated at a desired rate and the loss in the weight of the sample is recorded. Such changes in weight can be due to evaporation of moisture, evolution of gases, and chemical decomposition reactions, *i.e.* oxidation.

TGA is generally combined with DTA, and a plot of the loss in weight together with the DTA thermogram is recorded. These plots give information on the physical and chemical processes which are taking place. In an explosive reaction there is a rapid weight loss after ignition due to the production of gaseous substances and at the same time heat is generated. A typical DTA and TGA thermogram of HMX-β is shown in Figure 6.6.

The endotherm at 192 °C is due to the β–δ crystalline phase change, and the exotherm at 276 °C is due to the violent decomposition of HMX. Thermal pre-ignition and ignition temperatures of explosive substances can be obtained from DTA and TGA thermograms.

Differential Scanning Calorimetry

The technique of differential scanning calorimetry (DSC) is very similar to DTA. The peaks in a DTA thermogram represent a difference in temperature between the sample and reference, whereas the peaks in a DSC thermogram represent the amount of electrical energy supplied to the system to keep the sample and reference at the same temperature.

The areas under the DSC peaks will be proportional to the enthalpy change of the reaction.

DSC is often used for the study of equilibria, heat capacities and kinetics of explosive reactions in the absence of phase changes, whereas DTA combined with TGA is mainly used for thermal analysis.

Chapter 7

Manufacture of Explosives

NITRATION

Nitration plays an important role in the preparation of explosives. For example, the most commonly used military and commercial explosive compounds such as TNT, RDX, nitroglycerine, PETN, *etc.*, are all produced by nitration. Nitration is a chemical reaction by which nitro (NO_2) groups are introduced into organic compounds. It is basically a substitution or double exchange reaction in which one or more NO_2 groups of the nitrating agent replace one or more groups (usually hydrogen atoms) of the compound being nitrated. The nitration reaction can be classified into three categories as shown in Figure 7.1.

A summary of the nitration techniques for some military and commercial explosives is presented in Table 7.1.

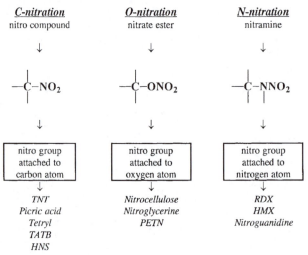

Figure 7.1 *Classification of explosive composition by nitration reaction*

118

Table 7.1 *Examples of nitrating agents for the manufacture of explosives*

Compound	Usual nitrating agent
C-Nitration	
Picric acid	Mixture of nitric and sulfuric acids
TNT	Mixture of nitric and sulfuric acids
HNS	Mixture of nitric and sulfuric acids
O-Nitration	
Nitroglycerine	Mixture of nitric and sulfuric acids
Nitrocellulose	Mixture of nitric and sulfuric acids
PETN	Mixture of nitric and sulfuric acids
N-Nitration	
Tetryl	Mixture of nitric and sulfuric acids
RDX	Nitric acid and ammonium nitrate
HMX	Nitric acid and ammonium nitrate

C-NITRATION

Picric Acid

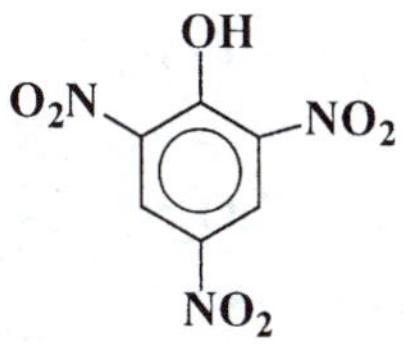

(7.1)

Picric acid (7.1) can be prepared by dissolving phenol in sulfuric acid and then nitrating the product with nitric acid as shown in Reaction 7.1.

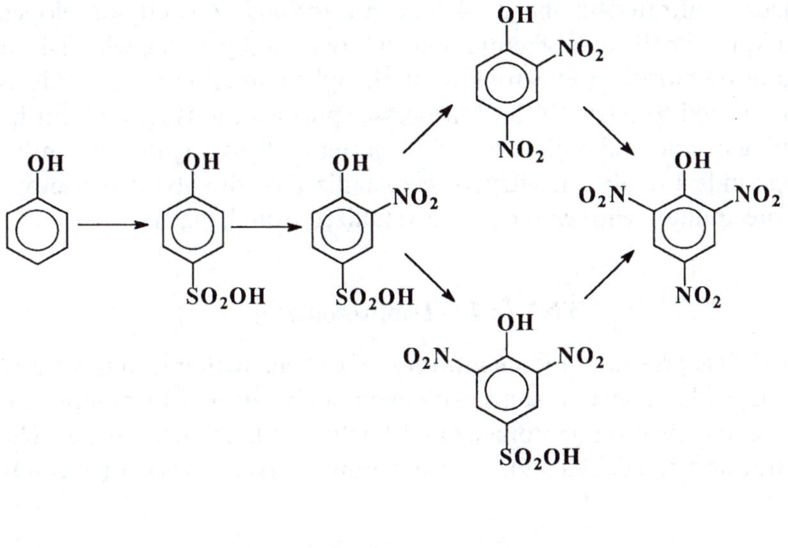

(7.1)

Sulfuric acid acts as an inhibitor or moderator of the nitration. The sulfonation of phenol at low and high temperatures produces the *ortho-* and *para*-sulfonic acids, respectively. All these substances yield picric acid as the final product of the nitration.

Tetryl

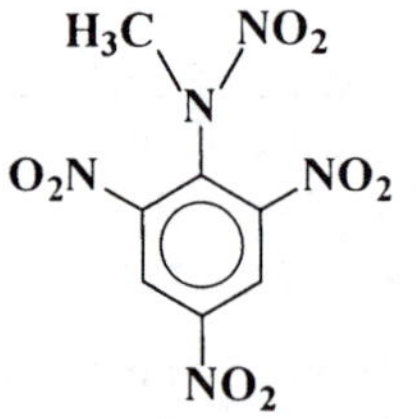

(7.2)

Tetryl (7.2) can be prepared by dissolving dimethylaniline in sulfuric acid and adding nitric and sulfuric acids at 70 °C as shown in Reaction 7.2.

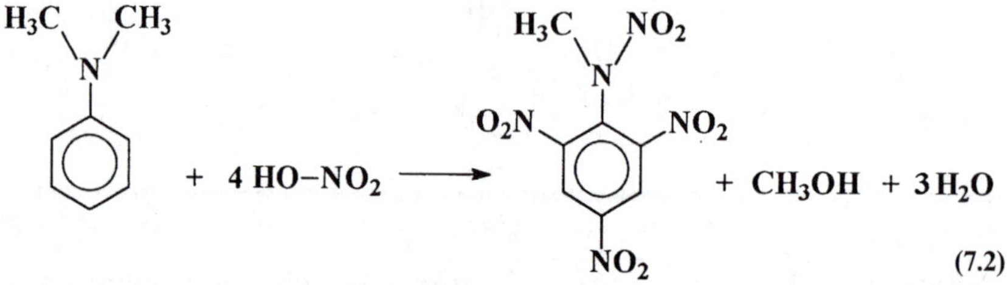

(7.2)

Here, one methyl group is oxidized and at the same time the benzene nucleus is nitrated in the 2-, 4- and 6-positions. Recently-developed techniques for the manufacture of tetryl treat methylamine with 2,4- or 2,6-dinitrochlorobenzene to give dinitrophenylmethylamine. This is then nitrated to tetryl. In both processes purification is carried out by washing in cold and boiling water, the latter hydrolysing the tetra-nitro compounds. Finally, the tetryl is recrystallized by dissolving in acetone and precipitated with water, or recrystallized from benzene.

TNT (2,4,6-Trinitrotoluene)

TNT (7.3) is produced by the nitration of toluene with mixed nitric and sulfuric acids in several steps. Toluene is first nitrated to mononitro-toluene and then dinitrotoluene and finally crude trinitrotoluene. The trinitration step needs a high concentration of mixed acids with free SO_3

groups. A summary for the preparation of trinitrotoluenes from toluene is presented in Reaction 7.3 (overleaf).

$$
\begin{array}{c}
CH_3 \\
O_2N \diagup\bigcirc\diagdown NO_2 \\
NO_2
\end{array}
$$

(7.3)

Nowadays, nitration of toluene is a continuous process, where toluene enters the reactor at one end and trinitrotoluene is produced at the other end. The nitrating acid flows in the opposite direction to the toluene and is topped up as required at various points. Stirring of the reactants plays an important part in the reaction since it speeds up the nitration process and helps to increase the yield.

Crude TNT contains isomers and nitrated phenolic compounds resulting from side reactions. The usual method of purification is to treat crude TNT with 4% sodium sulfite solution at pH 8–9, which converts the unsymmetrical trinitro compounds to sulfonic acid derivatives. These by-products are then removed by washing with an alkaline solution. Pure TNT is then washed with hot water, flaked and packed. It is important to remove the waste acid and unsymmetrical trinitrotoluenes together with any by-products of nitration as they will degrade the TNT, reduce its shelf life, increase its sensitivity and reduce its compatibility with metals and other materials. Trace amounts of unsymmetrical trinitrotoluenes and by-products will also lower the melting point of TNT. TNT can be further recrystallized from organic solvents or 62% nitric acid.

TATB (1,3,5-Triamino-2,4,6-trinitrobenzene)

TATB (7.4) is produced from the nitration of 1,3,5-trichloro-2,4,6-trinitrobenzene.

$$
\begin{array}{c}
NH_2 \\
O_2N \diagup\bigcirc\diagdown NO_2 \\
H_2N \qquad NH_2 \\
NO_2
\end{array}
$$

(7.4)

Reaction (7.3)

1,3,5-Trichloro-2,4,6-trinitrobenzene is prepared by the nitration of trichlorobenzene with a mixture of nitric acid and sulfuric acid. 1,3,5-Trichloro-2,4,6-trinitrobenzene is then converted to 1,3,5-trinitro-2,4,6-triaminobenzene (TATB) by nitrating with ammonia as shown in Reaction 7.4. The yellow-brown crystals of TATB are filtered and washed with water.

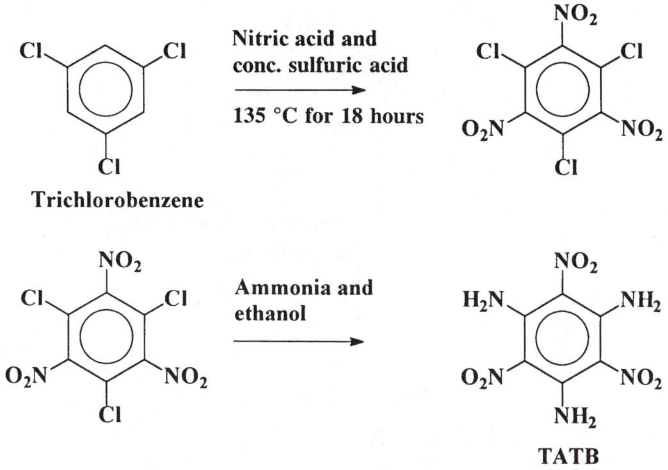

(7.4)

TATB is an explosive which is resistant to high temperatures, and therefore used in high-temperature environments or where safety from accidental fires is important. TATB is extremely insensitive to initiation by shock and requires a large amount of booster to initiate it. TATB is therefore regarded as an insensitive explosive and will most likely replace HMX and RDX in future explosive compositions. However, the cost of TATB is five to ten times greater than the cost of HMX.

HNS (Hexanitrostilbene)

HNS (7.5) can be prepared by many methods: these include the reaction of nitro derivatives of toluene with benzaldehyde, the reaction of nitro derivatives of benzyl halogenides with alkaline agents by removing hydrogen halogenide, and the oxidation of nitro derivatives of toluene.

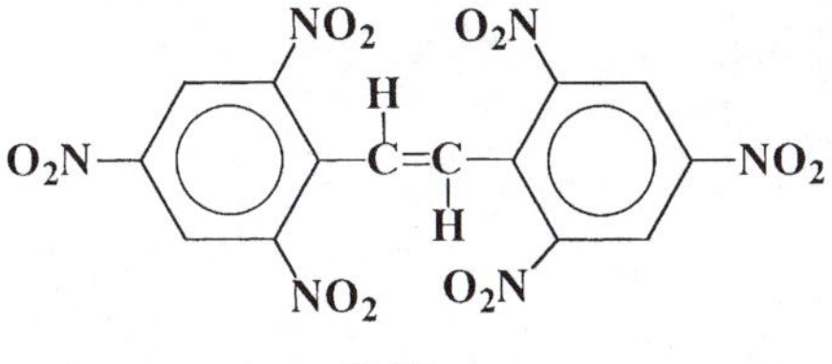

(7.5)

The first reaction involves heating together a mixture of trinitrotoluene and trinitrobenzaldehyde at temperatures of 160–170 °C, and then allowing the mixture to cool for two hours. The resultant product is a low yield of hexanitrostilbene (HNS). An increase in the yield of HNS can be achieved by reacting 2,4,6-trinitrobenzyl halogenide with potassium hydroxide in methanol as shown in Reaction 7.5.

(7.5)

HNS can also be prepared by the oxidation of TNT with sodium hypochlorite. Ten parts of 5% sodium hypochlorite solution are mixed with a chilled solution of one part TNT in ten parts methanol. The solution is allowed to stand at ambient temperature until HNS precipitates as a fine crystalline product. HNS is then recrystallized from nitrobenzene to give pale yellow-coloured needles. The mechanism of the reaction is presented in Reaction 7.6.

HNS is often used in military explosive compositions as a crystal modifier for TNT. Around $\frac{1}{2}$–1% of HNS is added to molten TNT so that on cooling, small, randomly-orientated crystals are formed. These small crystals have a high tensile strength and prevent the TNT composition from cracking.

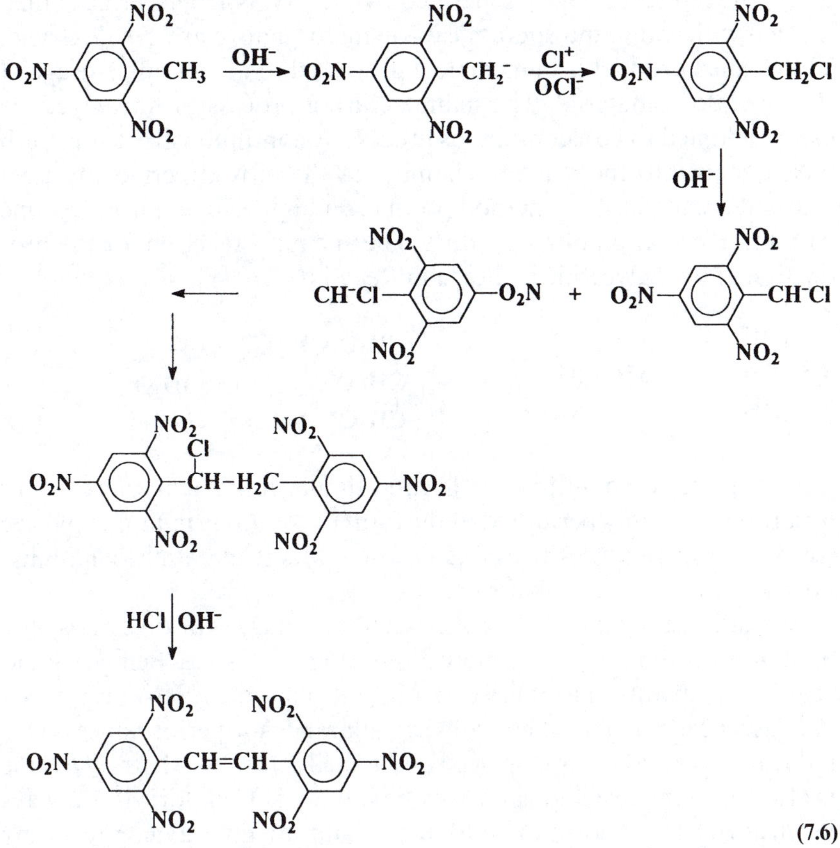

NO₂ … (reaction scheme 7.6)

$$O_2N-\!\!\bigcirc\!\!-CH_3 \xrightarrow{OH^-} O_2N-\!\!\bigcirc\!\!-CH_2^- \xrightarrow[OCl^-]{Cl^+} O_2N-\!\!\bigcirc\!\!-CH_2Cl$$

$$\xrightarrow{OH^-}$$

$$CHCl-\!\!\bigcirc\!\!-O_2N \;+\; O_2N-\!\!\bigcirc\!\!-CHCl$$

$$O_2N-\!\!\bigcirc\!\!-\overset{Cl}{\underset{}{CH}}-H_2C-\!\!\bigcirc\!\!-NO_2$$

$$\xrightarrow{HCl \mid OH^-}$$

$$O_2N-\!\!\bigcirc\!\!-CH=CH-\!\!\bigcirc\!\!-NO_2$$

(7.6)

O-NITRATION

Nitroglycerine

Nitroglycerine (7.6) is prepared by injecting highly-concentrated glycer-ine into a mixture of highly-concentrated nitric and sulfuric acids at a controlled temperature.

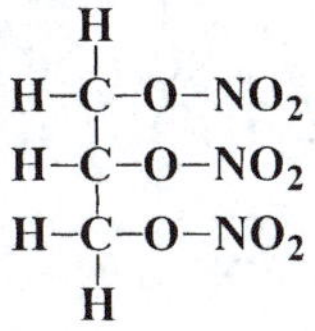

$$\begin{array}{c} H \\ H-C-O-NO_2 \\ H-C-O-NO_2 \\ H-C-O-NO_2 \\ H \end{array}$$

(7.6)

The mixtures are constantly stirred and cooled with brine. At the end of the reaction the nitroglycerine and acids are poured into a separator, where the nitroglycerine is separated by gravity. Nitroglycerine is then washed with water and sodium carbonate to remove any residual acid. The cleaned product is immediately processed to avoid bulk storage of the hazardous material. The manufacturing process of nitroglycerine can be adapted to be a continuous process by continuously feeding fresh mixed acids into the reaction chamber as the nitroglycerine and used acids are removed. This method produces a high yield of nitroglycerine with little formation of by-product. The chemical reaction for the production of nitroglycerine is shown in Reaction 7.7.

$$
\begin{array}{l}
CH_2OH \\
CHOH \\
CH_2OH
\end{array}
\quad + \quad 3\,HONO_2 \quad \longrightarrow \quad
\begin{array}{l}
CH_2ONO_2 \\
CHONO_2 \\
CH_2ONO_2
\end{array}
\quad + \quad 3H_2O
\tag{7.7}
$$

Transportation of nitroglycerine and similar nitric acid esters is very hazardous and only permitted in the form of solutions in non-explosive solvents or as mixtures with fine-powdered inert materials containing not more that 5% nitroglycerine.

Nowadays, nitroglycerine is not used for military explosives; it is only used in propellants and commercial blasting explosives. Being a liquid at room temperature it pours easily into its container. However, when used as a propellant it suffers from irregular ballistic performances. This is due to the liquid being displaced as the shell is rotating. Nitroglycerine is also very sensitive and can easily be initiated. Therefore, it is always desensitized by absorption into Kieselguhr to give dynamite or by gelling with nitrocellulose to produce commercial blasting gels.

Nitrocellulose

(7.7)

The manufacturing method used today for the nitration of cellulose employs a mixture of sulfuric and nitric acids. During nitration of the cellulose some of the —OH groups in the molecules are replaced by the nitrate groups —ONO$_2$ as shown in Reaction 7.8.

$$6[HO-NO_2]_n$$

$$+ \ 6[H_2O]_n$$

(7.8)

The reaction is able to take place without destroying the fibrous structure of the cellulose. The extent of nitration of the —OH groups is dependent on the reaction conditions, particularly the composition of the acid, the time and temperature of nitration.

Cotton, cotton linters (short fibres) or wood cellulose are cleaned by mechanical combing, and then bleached to give a more open structure for nitration. The fibres are mixed with the nitrating acid in a pre-nitrator and passed into a second stage of nitration where very close contact is obtained between the fibres and the nitrating acid. The nitrated fibres then pass into a third stage where nitration is completed. The nitrating acid is removed from nitrocellulose by centrifugation. For low-nitrated products the nitrocellulose is washed with dilute acid and water, whereas high-nitrated products continue on to the next stage without being washed. Nitrocellulose is then mixed with water and boiled under high pressures and turned into a pulp by squeezing the wet nitrocellulose fibres between rollers which contain knife blades. The pulped nitrocellulose is then boiled, washed, cleaned to remove foreign particles, and packed. A flow diagram for the manufacturing process of nitrocellulose is presented in Figure 7.2.

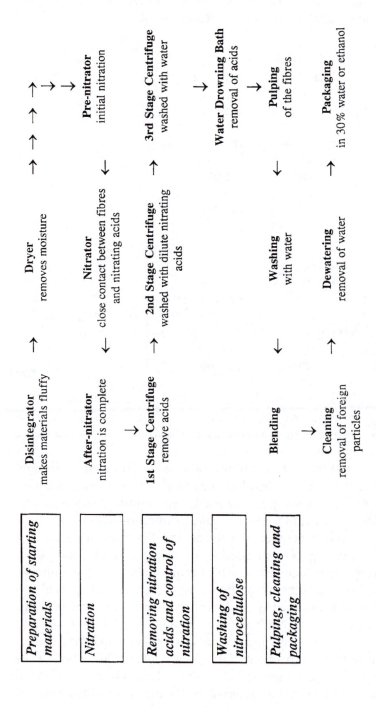

Figure 7.2 *Flow diagram for the manufacturing process of nitrocellulose*

Nitrocellulose can be quite hazardous if left to dry out completely; therefore, it is usually stored and transported in 30% water or ethanol. Nitrocellulose is often dissolved in solvents to form a gel. For example, commercial explosives used for blasting purposes contain nitrocellulose dissolved in nitroglycerine, and some gun propellant compositions contain nitrocellulose dissolved in a mixture of acetone and water.

The degree of nitration plays a very important role in the application of nitrocellulose. Guncotton contains 13.45% nitrogen and is used in the manufacture of double-base and high energy propellants, whereas nitrocellulose used in commercial gelatine and semi-gelatine dynamite contains 12.2% nitrogen.

PETN (Pentaerythritol tetranitrate)

PETN (7.8) is prepared by nitrating pentaerythritol.

$$O_2N-O-H_2C \quad CH_2-O-NO_2$$
$$C$$
$$O_2N-O-H_2C \quad CH_2-O-NO_2$$

(7.8)

Pentaerythritol is made by mixing formaldehyde with calcium hydroxide in an aqueous solution held at 65–70 °C. Nitration of pentaerythritol can be achieved by adding it to concentrated nitric acid at 25–30 °C to form PETN. The crude PETN is removed by filtration, washed with water, neutralized with sodium carbonate solution and recrystallized from acetone. This manufacturing process for PETN results in 95% yield with negligible by-products. The process is summarized in Reaction 7.9 (overleaf).

PETN is used in detonating cords and demolition detonators. In detonating cords PETN is loaded into plastic tubes. PETN is also mixed with TNT and a small amount of wax to form 'Pentolite' mixtures. Pentolite mixtures are used as booster charges for commercial blasting operations.

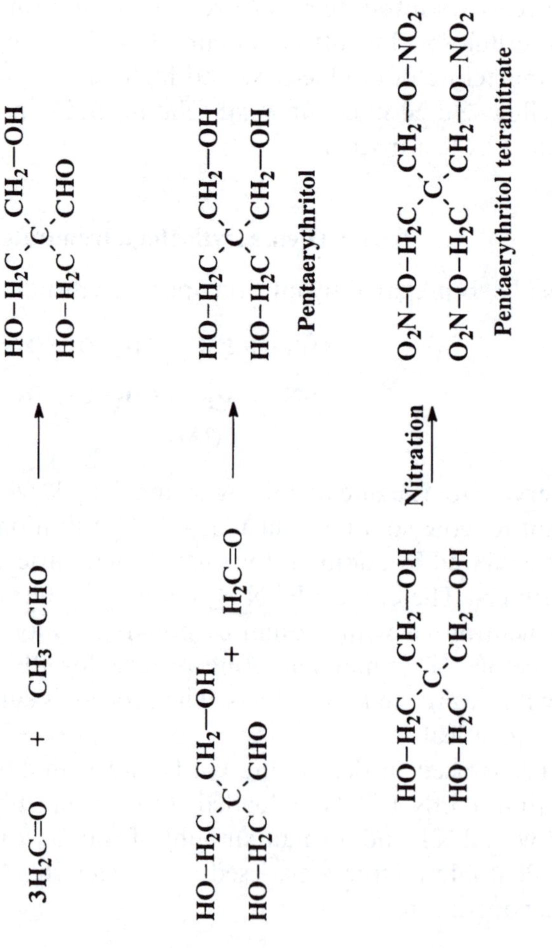

$$3H_2C=O \quad + \quad CH_3-CHO \quad \longrightarrow \quad \begin{array}{c} HO-H_2C \\ HO-H_2C \end{array}\!\!> C <\!\!\begin{array}{c} CH_2-OH \\ CHO \end{array}$$

$$\begin{array}{c} HO-H_2C \\ HO-H_2C \end{array}\!\!> C <\!\!\begin{array}{c} CH_2-OH \\ CHO \end{array} \quad + \quad H_2C=O \quad \longrightarrow \quad \begin{array}{c} HO-H_2C \\ HO-H_2C \end{array}\!\!> C <\!\!\begin{array}{c} CH_2-OH \\ CH_2-OH \end{array}$$

Pentaerythritol

$$\begin{array}{c} HO-H_2C \\ HO-H_2C \end{array}\!\!> C <\!\!\begin{array}{c} CH_2-OH \\ CH_2-OH \end{array} \quad \xrightarrow{\text{Nitration}} \quad \begin{array}{c} O_2N-O-H_2C \\ O_2N-O-H_2C \end{array}\!\!> C <\!\!\begin{array}{c} CH_2-O-NO_2 \\ CH_2-O-NO_2 \end{array}$$

Pentaerythritol tetranitrate

Reaction (7.9)

N-NITRATION

RDX (Cyclotrimethylenetrinitramine)

The simplest method of preparing RDX (7.9) is by adding hexamethylenetetramine to excess concentrated nitric acid at 25 °C and warming to 55 °C. RDX is precipitated with cold water and the mixture is then boiled to remove any soluble impurities. Purification of RDX is carried out by recrystallization from acetone.

(7.9)

The chemistry of the preparation of RDX is highly complex. When hexamethylenetetramine is reacted with nitric acid, hexamethylenetetramine dinitrate is formed which is then nitrated to form an intermediate 'Compound I' as shown in Reaction 7.10.

Hexamethylenetetramine   Hexamethylenetetramine dinitrate

Hexamethylenetetramine dinitrate   Compound I   (7.10)

Further nitrolysis of Compound I results in the formation of RDX via two other intermediate compounds as shown in Reaction 7.11.

Other manufacturing techniques include the SH-process invented by Schnurr, the K-process patented by Knöffler, the W-process patented by Wolfram, the E-process patented by Erbele, and the KA-method invented by Knöffler and Apel and also Bachmann.

The SH-process involves continuous nitration of hexamethylenetetramine by concentrated nitric acid, with the production of nitrous gas. The RDX is filtered from the residual acid and stabilized by boiling in water under pressure and purified by recrystallization from acetone.

In the K-process RDX is formed by reacting ammonium nitrate with a mixture of hexamethylenetetramine and nitric acid, and warmed as shown in Reaction 7.12.

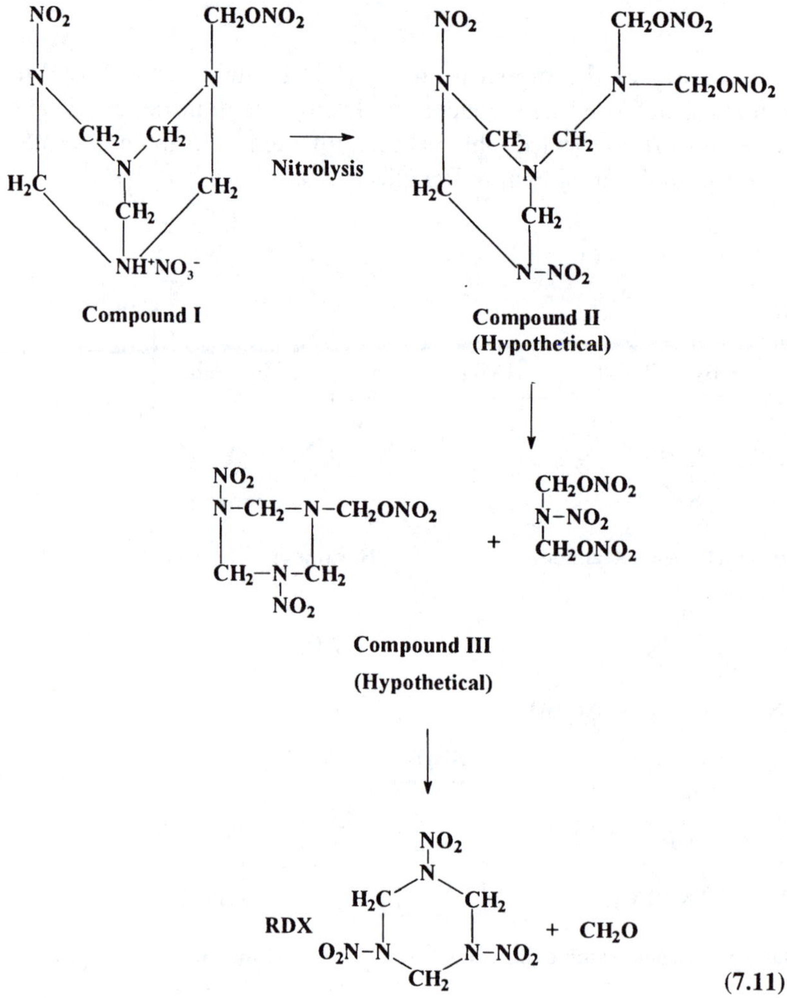

Compound I

$\xrightarrow{\text{Nitrolysis}}$

Compound II
(Hypothetical)

Compound III
(Hypothetical)

RDX   + CH_2O

(7.11)

$$+ \quad 2NH_4NO_3 \quad + \quad 4HNO_3$$

$$2 \quad \left[\begin{array}{c} \end{array} \right] \quad + \quad 6H_2O$$

(7.12)

The W-process is based on the condensation of potassium amido-sulfonate with formaldehyde, and the nitration of the condensation product as shown in Reaction 7.13 (overleaf). Potassium amidosulfon-ate and formaldehyde are reacted together to produce potassium methyleneamidosulfonate. This product is nitrated to RDX by a mix-ture of nitric and sulfuric acids.

In the E-process paraformaldehyde and ammonium nitrate undergo dehydration by acetic anhydride solution resulting in the formation of RDX as shown in Reaction 7.14.

The by-product formaldehyde can be transformed into hexa-methylenetetramine by reacting it with ammonium nitrate and thus increasing the yield.

Lastly, the KA-process (Bachmann process) is based on the reaction between hexamethylenetetramine dinitrate and ammonium nitrate with a small amount of nitric acid in an acetic anhydride solution. The chemical reaction is presented in Reaction 7.15.

RDX has a melting point close to its ignition temperature and there-fore cannot be safely melted. However, it is often used in conjunction with TNT for casting techniques. RDX is very sensitive and cannot be processed by pressing. In order to reduce its sensitivity a binder such as a wax or polymer is used, these binders are known as phlegmatizing binders. RDX is the main explosive component of British explosive compositions. It is used in shells, bombs, shaped charges, exploders, detonators and polymer bonded explosives.

$$\underset{\substack{\text{Potassium} \\ \text{amidosulfonate}}}{\overset{\displaystyle \underset{\;}{\overset{\text{OK}}{\underset{\text{SO}_2}{|}}}}{\underset{\text{NH}_2}{}}} \; + \; \underset{\text{Formaldehyde}}{\text{CH}_2\text{O}} \longrightarrow \underset{\substack{\text{Potassium} \\ \text{methyleneamidosulfonate}}}{\overset{\text{OK} \atop \text{SO}_2}{\underset{\text{N=CH}_2}{}}}$$

$$3\,\underset{\substack{\text{Potassium} \\ \text{methyleneamidosulfonate}}}{\overset{\text{OK} \atop \text{SO}_2}{\underset{\text{N=CH}_2}{}}} \; + \; \underset{\substack{\text{Nitric} \\ \text{acid}}}{3\text{HNO}_3} \longrightarrow \quad \text{RDX} \; + \; 3\text{KHSO}_4$$

RDX ring structure: $O_2N\text{–}N$, CH_2, $N\text{–}NO_2$, $NO_2\text{–}N$, H_2C, CH_2

Reaction (7.13)

$$6CH_2O + 4NH_4NO_3 + 3(CH_3CO)_2O \longrightarrow C_6H_{12}N_4 + 4HNO_3 + 6CH_3COOH + 3H_2O$$

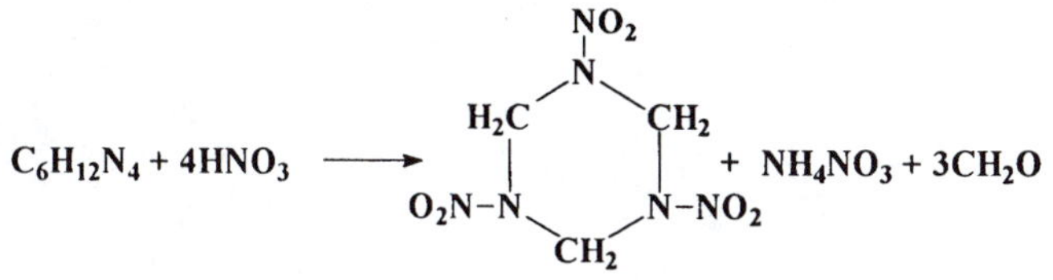

(7.14)

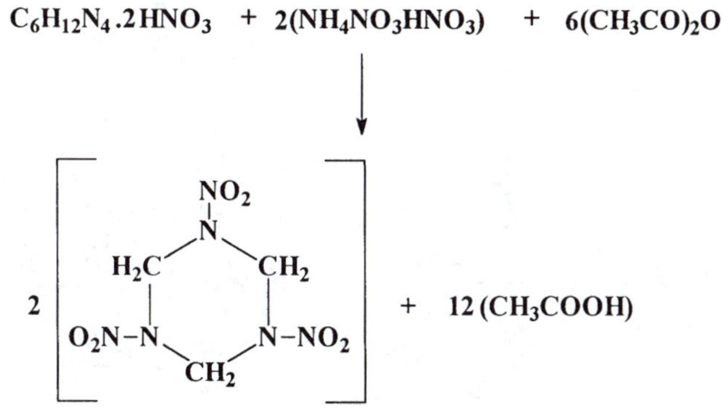

(7.15)

HMX (Cyclotetramethylenetetranitramine)

HMX (7.10) is formed as a by-product during the manufacture of RDX by the Bachman process.

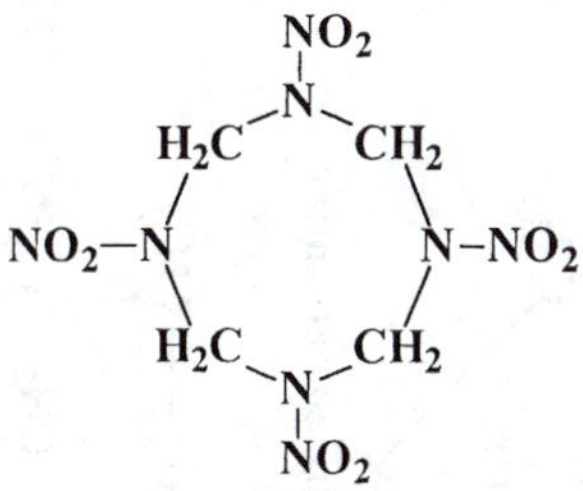

(7.10)

Hexamethylenetetramine, acetic acid, acetic anhydride, ammonium nitrate and nitric acid are mixed together and held at 45°C for 15 min. Ammonium nitrate, nitric acid and acetic anhydride are then slowly added and left on a steam bath for 12 h. A precipitate forms containing 27% RDX and 73% HMX. This process is shown in Reaction 7.16 (overleaf).

RDX is destroyed by placing the precipitate into a hot, aqueous solution of sodium tetraborate decahydrate with a small amount of

Reaction (7.16)

sodium hydroxide. The RDX is completely destroyed when the value of the pH increases to greater than 9.7. HMX is filtered and recrystallized from nitromethane to give the β-form of HMX.

HMX is similar to RDX in that its melting point is very close to its ignition temperature and therefore cannot be safely melted. It can be used with TNT for casting techniques. HMX is also very sensitive and is used with phlegmatizing binders. HMX has a slightly higher density and melting temperature than RDX and is overall a better explosive with respect to its performance. However its cost is approximately three times greater than RDX.

Nitroguanidine

Nitroguanidine (7.11) exists in at least two crystalline forms, the α- and β-forms.

$$NH=C\begin{array}{c} NH_2 \\ NH-NO_2 \end{array}$$

(7.11)

The α-form may be prepared by dissolving guanidine nitrate in concentrated sulfuric acid and pouring into excess water before crystallizing from hot water. Long, thin, flexible, lustrous needles, which are very tough, are formed. This is the most common form and is used in the explosive industry. The β-form may be prepared by nitrating a mixture of guanidine sulfate and ammonium sulfate and crystallizing from hot water to form fern-like clusters of small, thin, elongated plates. The β-form may be converted into the α-form by dissolving in concentrated sulfuric acid and immersing in excess water. In order to reduce the crystal size so that it can be incorporated into colloidal propellants, a hot solution of nitroguanidine is sprayed on to a cooled metallic surface and allowed to cool in a stream of cold air. The resultant nitroguanidine is a fine powder. The reaction scheme for the preparation of nitroguanidine is presented in Reaction 7.17.

$$\begin{array}{c} H_2N \\ H_2N \end{array}C=NH \ + \ HO-NO_2 \ \longrightarrow \ \begin{array}{c} H_2N \\ H_2N \end{array}C=N-NO_2 \ + \ H_2O$$

(conc. H_2SO_4)

(7.17)

Ammonium Nitrate

The most common method of manufacturing ammonium nitrate is by injecting gaseous ammonia into 40–60% nitric acid at 150 °C as shown in Reaction 7.18:

$$NH_3 + HNO_3 \rightarrow NH_4NO_3 \qquad (7.18)$$

Dense ammonium nitrate crystals are formed by spraying droplets of molten ammonium nitrate solution (>99.6%) down a short tower. The spray produces spherical particles known as 'prills'. These crystals are non-absorbent and used in conjunction with nitroglycerine. An absorbent form of ammonium nitrate can be obtained by spraying a hot, 95% solution of ammonium nitrate down a high tower. The resultant spheres are carefully dried and cooled to prevent breakage during handling. These absorbent spheres are used with fuel oil.

Ammonium nitrate is the cheapest source of oxygen available for commercial explosives. It is used by itself, in conjunction with fuels, or with other explosives such as nitroglycerine and TNT.

PRIMARY EXPLOSIVES

Manufacture of primary explosives is very hazardous and accidents such as explosions can occur during the preparation. Therefore strict safety procedures are always adhered to.

Lead Azide

Lead azide (7.12) is prepared by dissolving lead nitrate in a solution containing dextrin, with the pH adjusted to 5 by adding one or two drops of sodium hydroxide.

$$
\begin{array}{c}
N{=}N^+{=}N^- \\
/ \\
Pb \\
\diagdown \\
N{=}N^+{=}N^-
\end{array}
$$

(7.12)

This solution is heated to 60–65 °C and stirred. Sodium azide dissolved in a solution of sodium hydroxide is then added dropwise to the lead nitrate solution. The mixture is then left to cool to room temperature with continuous stirring. Lead azide crystals are filtered, washed with water and dried. This process is presented in Reaction 7.19.

$$2Na(N_3)_{2(aq)} + Pb(NO_3)_{2(aq)} \rightarrow Pb(N_3)_{2(s)} + 2NaNO_{3(aq)} \qquad (7.19)$$

Lead azide crystals should be spherical in shape, opaque in appearance and less than 0.07 mm in diameter. Dextrin is added as a colloiding agent, which prevents the formation of large, sensitive crystals of lead azide and regulates, to some extent, the shape of the crystals.

Mercury Fulminate

Mercury fulminate is prepared by dissolving mercury in nitric acid and then pouring into ethanol. A vigorous reaction takes pace which is accompanied by the evolution of white fumes, then by brownish-red fumes and finally again by white fumes. At the same time crystals of mercury fulminate are formed. The crystals are recovered and washed with water until all of the acid is removed.

$$(C\equiv NO)_2Hg$$

(7.13)

The crystals are a greyish colour and are stored under water. The mechanism for this reaction and the intermediate steps are presented in Reaction 7.20.

1. Oxidation of ethanol to ethanal
$$CH_3CH_2OH + HNO_3 \rightarrow CH_3CHO + HNO_2 + H_2O$$

2. Formation of nitrosoethanal (nitrosation)
$$CH_3CHO + HNO_2 \rightarrow NOCH_2CHO + H_2O$$

3. Isomerization of nitrosoethanal to isonitrosoethanal
$$NOCH_2CHO \rightarrow HON=CH-CHO$$

4. Oxidation of isonitrosoethanal to isonitrosoethanoic acid
$$HON=CH-CHO \rightarrow HON=CH-COOH$$

5. Decomposition of isonitrosoethanoic acid to fulminic acid and
   methanoic acid
$$HON=CH-COOH \rightarrow C\equiv NOH + HCOOH$$

6. Formation of mercury fulminate
$$2C\equiv NOH + Hg(NO_3)_2 \rightarrow (C\equiv NO)_2Hg + 2HNO_3$$

(7.20)

Tetrazene

Tetrazene (7.14) is prepared by dissolving sodium nitrite in distilled water and warming to 50–55 °C. An acidic solution of aminoguanidine sulfate is then added and the mixture stirred for 30 min. A precipitate of tetrazene forms, which is decanted, washed with water and then alcohol, and dried at 45–50 °C.

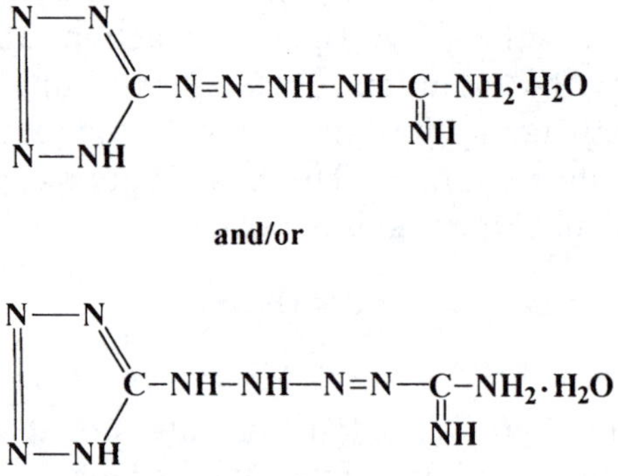

$$N—N \\ \diagdown \qquad \diagdown \\ \diagup \qquad \diagup \\ N—NH$$

C–N=N–NH–NH–C–NH$_2$·H$_2$O
                          ‖
                          NH

and/or

$$N—N \\ \diagdown \qquad \diagdown \\ \diagup \qquad \diagup \\ N—NH$$

C–NH–NH—N=N—C–NH$_2$·H$_2$O
                          ‖
                          NH

(7.14)

The mechanism for this reaction is presented in Reaction 7.21.

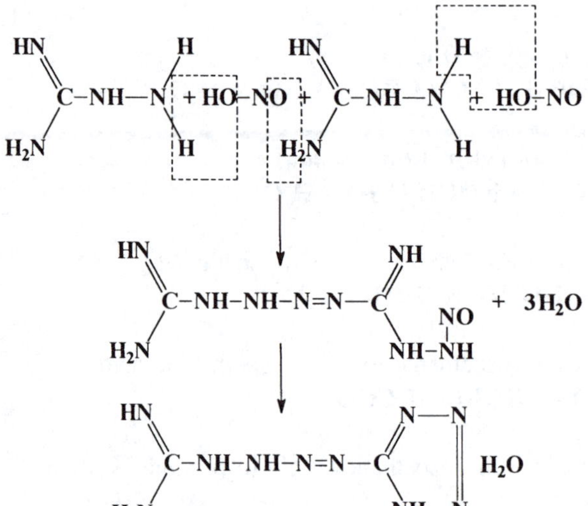

HN
 ‖
 C–NH—N + HO–NO + C–NH—N + HO–NO
 | | | |
H$_2$N   H   H$_2$N   H

HN   NH
 ‖   ‖
 C–NH–NH–N=N—C   + 3H$_2$O
 | NO
H$_2$N   NH–NH

HN   N—N
 ‖   ‖
 C–NH–NH–N=N—C   H$_2$O
 | NH—N
H$_2$N

(7.21)

The rate at which the aminoguanidine sulfate is added to the solution of sodium nitrite determines the size of the tetrazene crystals. If the aminoguanidine sulfate is added quickly, small crystals are formed, whereas large crystals are formed when the solution is introduced slowly. Dextrin can be added to the reacting solution to obtain a more uniform crystal size.

COMMERCIAL EXPLOSIVE COMPOSITIONS

Ammonium Nitrate

Ammonium nitrate based explosives are generally used for quarrying, tunnelling and mining. They are mixtures of ammonium nitrate, carbon carriers such as wood meal, oils or coal, and sensitizers such as nitroglycol, TNT and dinitrotoluene. These compositions may also contain aluminium powder to improve their performance.

Ammonium nitrate prills are often mixed with fuel oil (liquid hydrocarbons) to produce a commercial explosive mixture known as 'ANFO' which is used in quarrying. ANFO can be prepared in a factory by mixing both ingredients in a rotating container and dispensing the product into polyethylene or cardboard tubes. The tubes are then sealed and transported to the place of use. ANFO can also be prepared at the site where the explosive composition is to be used. Fuel oil is poured into a polyethylene bag containing ammonium nitrate and left for some time to allow the oil to soak into the ammonium nitrate. The ANFO mixture is then poured from the polyethylene bag into the hole (*i.e.* shot-hole) where the explosive mixture is detonated.

ANFO can be mixed directly in the shot-hole by first pouring the ammonium nitrate into the shot-hole followed by the fuel oil. The main advantage of mixing on site is that no safety procedures are required for the transportation of fuel oil and ammonium nitrate, since fuel oil and ammonium nitrate are not classed as explosives. It is only when they are mixed together that the composition becomes an explosive substance.

Higher density and improved water-resistant ammonium nitrate compositions can be obtained by mixing ammonium nitrate with 20–40% gelatinized nitroglycol or gelatinized nitroglycerine and nitroglycol mixtures. Ammonium nitrate and TNT/dinitrotoluene mixtures also have good water-resistant properties since the TNT/dinitrotoluene forms a coating around the ammonium nitrate crystals.

Ammonium Nitrate Slurries

Ammonium nitrate slurries can be either prepared in the factory and loaded into cartridges or mixed on site and pumped down the shot-holes. Ammonium nitrate slurries consist of a saturated aqueous solution of ammonium (about 65%) and other nitrates (*i.e.* methylamine nitrate 'MAN'). This solution also contains additional amounts of un-dissolved nitrates together with a fuel. The fuel is generally aluminium powder but water-soluble fuels such as glycol may also be employed. The slurries can be made more sensitive by adding either TNT, PETN,

Table 7.2 *Summary of dynamite compositions*

Name	Fuel	Oxidizer
Gelatine dynamite		
Gelatine dynamite	25–55% Nitroglycerine, 1–5% nitrocellulose, woodmeal	Inorganic nitrates
Semi-gelatine dynamite	15–20% Nitroglycerine, 1–5% nitrocellulose, woodmeal	Inorganic nitrates
Gelignite	25–55% Nitroglycerine, 1–5% nitrocellulose, woodmeal	Sodium or potassium nitrate
Ammonia gelignite	25–55% Nitroglycerine, 1–5% nitrocellulose, woodmeal	Ammonium nitrate
Non-gelatine dynamite		
Non-gelatine dynamite	10–50% Nitroglycerine, woodmeal	Sodium or potassium nitrate
Ammonia dynamite	10–50% Nitroglycerine, woodmeal	Ammonium nitrate

etc., or by introducing finely-dispersed air bubbles in the form of air-filled microballoons.

Ammonium Nitrate Emulsion Slurries

Ammonium nitrate emulsion slurries contain two immiscible phases; the aqueous phase which supplies the majority of the oxygen for the reaction and the oil phase which supplies the fuel. Emulsion slurries are based on a water-in-oil emulsion, which is formed from a saturated aqueous solution of ammonium nitrate in a mineral oil phase. Emulsion slurries can be made more sensitive by adding air-filled microballoons. The density of the emulsion slurries is slightly higher than water gels, therefore resulting in a greater performance. Emulsion slurries can also be prepared in the factory and loaded into cartridges or mixed on site and pumped down the shot-holes.

Dynamite

Dynamite is a generic name for a variety of explosive compositions. These compositions can be divided into two categories; gelatine dynamite and non-gelatine dynamite, as shown in Table 7.2.

Gelatine dynamite is prepared by dissolving nitrocellulose in nitro-glycerine at 45–50 °C to form a gel. The mixture is continuously stirred by large, vertical mixer blades similar to a bread-making machine. Once the gel is formed the other ingredients are added. The explosive mixture is then extruded or pressed into long rods which are cut into smaller pieces and packaged into paper cartridges coated with paraffin. The manufacture of non-gelatine dynamite is similar to gelatine dynamite except that nitrocellulose is not used in the formulations.

MILITARY EXPLOSIVE COMPOSITIONS

Most military explosives are solid compounds which are manufactured in granular form, with bulk densities of less than 1 g cm^{-3}. These granular compounds are then mixed with other explosive or inert additives to give explosive compositions with densities between 1.5 and 1.7 g cm^{-3}. The explosive compositions are then cast, pressed or extruded into their final form.

Casting

The technique of casting is used for loading explosive compositions into large containers. Casting essentially involves heating the explosive composition until it melts, pouring it into a container and leaving it to solidify by cooling. Although the process sounds simple it has taken years of development to optimize the processing conditions. The disadvantages of casting are that the density of the cast compositions is not as high as pressed or extruded compositions, that cracks can sometimes occur when the composition shrinks on solidification resulting in a composition which is more sensitive, and that the solid ingredients tend to settle during solidification resulting in inhomogeneity. Explosive compositions which have ignition temperatures near to their melting temperatures cannot be cast. The advantages of casting are that the process is simple, cheap, flexible and produces high production volumes.

Explosive compositions which are processed by casting generally contain TNT, which has a relatively low melting temperature (80 °C) compared with its ignition temperature (240 °C). Large quantities of TNT were used for casting in the Second World War. Molten TNT was mixed with ammonium nitrate to give 'amatol', or ammonium nitrate and aluminium to give 'minol'. Today, TNT is used as an energetic binder for cast compositions. It is used to bind together RDX, HMX, aluminium and ammonium perchlorate.

The manufacturing process for the casting of explosive compositions nowadays uses carefully controlled cooling, vibration and vacuum techniques to produce a homogenous, solid composition. The purity and size of the explosive crystals are carefully monitored and the complete product is subjected to X-ray analysis for the detection of cracks.

Pressing

The technique of pressing is often used for loading powdered explosive compositions into small containers. Pressing does not require very high temperatures and can be carried out under vacuum. The whole process can be automated resulting in high volume productions. However, the machinery is expensive and the process is more hazardous than casting.

The explosive compositions can be pressed directly into a container or mould, and ejected as pellets as shown in Figure 7.3.

Variations in the density of the pressed compositions do occur, particularly near to the surface resulting in an anisotropic product. Pressing in incremental stages or using two pistons can reduce these variations as shown in Figures 7.4 and 7.5, respectively.

When dimensional stability, uniformity and high density are essential to the performance of the fabricated explosive composition, hydrostatic and isostatic pressing can be employed. In both cases the explosive composition is compressed by the action of a fluid instead of a piston. In hydrostatic pressing the explosive composition is placed on a solid surface and covered with a rubber diaphragm as shown in Figure 7.6. In isostatic pressing the composition is placed in a rubber bag which is then immersed into a pressurizable fluid as shown in Figure 7.7.

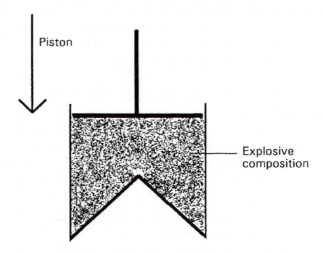

Figure 7.3 *Schematic diagram of pressing using a single piston*

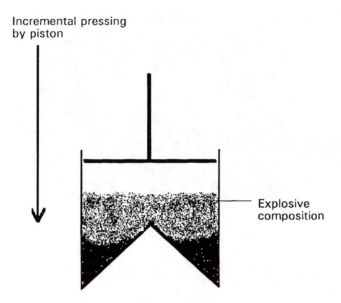

Figure 7.4 *Schematic diagram of incremental pressing*

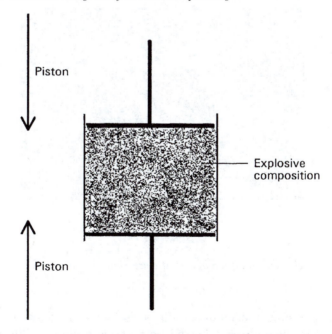

Figure 7.5 *Schematic diagram of pressing using two pistons*

Hydrostatic and isostatic pressing are also used to consolidate explosive compositions which are very sensitive to friction.

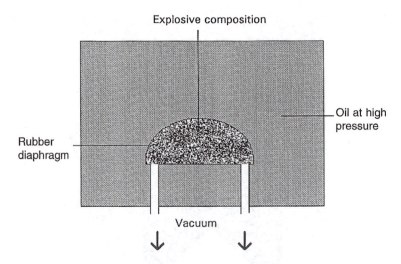

Figure 7.6 *Schematic diagram of hydrostatic pressing*

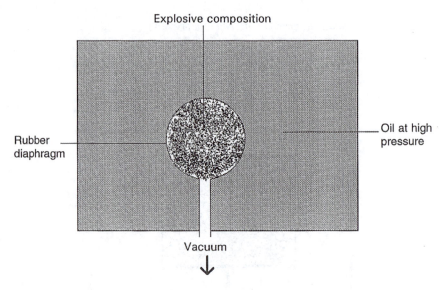

Figure 7.7 *Schematic diagram of isostatic pressing*

After pressing, the composition is cut into the desired shape by machining. This can be achieved by turning, milling, drilling, sawing and boring. The technique of machining an explosive composition is very hazardous and therefore is automated and carried out remotely.

Ram and Screw Extrusion

Manufacture of propellant compositions by ram extrusion has been in operation for the last two decades. The propellant composition is loaded into a barrel and extruded through a small hole under high pressures as shown in Figure 7.8.

The disadvantage of ram extrusion is that it is a batch process and requires mixing of the components beforehand. On the other hand, screw extrusion is a continuous process.

Screw extruders have been used for many years in the plastics industry and have recently been developed to mix and extrude explosive components which contain a polymeric binder, *i.e.* polymer bonded explosives (PBXs). A screw extruder is very similar to a meat-mincing machine where the powdered or granulated components go in at one end and are pushed along the barrel by a rotating screw. For explosive compositions the extruder generally has two rotating screws which are capable of mixing and compacting the explosive components as they travel along the barrel. A schematic diagram of a twin screw extruder is presented in Figure 7.9.

Extruded explosive compositions contain a polymer which binds the explosive crystals together. The polymer acts as an adhesive and helps the explosive composition to retain its shape after extrusion.

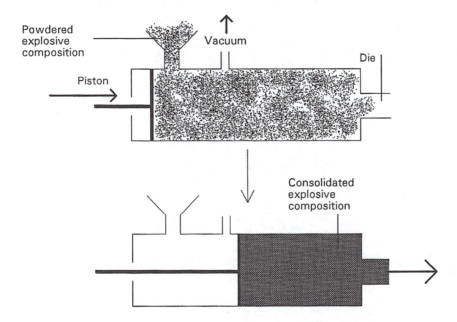

Figure 7.8 *Schematic diagram of ram extrusion*

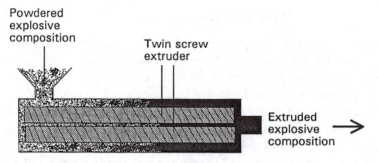

Figure 7.9 *Schematic diagram of twin screw extrusion*

Chapter 8

Introduction to Propellants and Pyrotechnics

INTRODUCTION TO PROPELLANTS

A propellant is an explosive material which undergoes rapid and pre-
dictable combustion (without detonation) resulting in a large volume of
hot gas. This gas can be used to propel a projectile, *i.e.* a bullet or a
missile, or in gas generators to drive a turbine, *i.e.* torpedoes.

In order to produce gas quickly a propellant, like a high explosive,
must carry its own oxygen together with suitable quantities of fuel
elements, *i.e.* carbon, hydrogen, *etc.* A homogeneous propellant (monopropellant for liquid propellants) is where the fuel and oxidizer are in the
same molecule, *i.e.* nitrocellulose, whereas a heterogeneous propellant
(bipropellant for liquid propellants) has the fuel and oxidizer in separate
compounds. Gun propellants are traditionally known to be homogeneous, whereas rocket propellants are heterogeneous.

GUN PROPELLANTS

Performance

Gun propellants are designed to provide large quantities of gas which is
used to propel projectiles at high kinetic energies. The velocity of the
projectile is dependent on the rate at which the gas is produced, which in
turn is dependent on the amount of chemical energy released, and the
efficiency of the gun η as shown in Equation 8.1.

$$mQ\eta = \frac{Mv^2}{2}$$

Therefore,
$$v = \sqrt{\frac{2mQ\eta}{M}} \tag{8.1}$$

Here, v is the muzzle velocity of the projectile in m s^{-1}, m and M are the mass of the propellant and projectile in grams, respectively, and Q is the amount of chemical energy released by the combustion of the propellants in J g^{-1}. The value for Q (also known as the 'Heat of Deflagration') can be calculated from the standard heats of formation as shown in Equation 8.2.

$$Q = \Sigma\Delta U^{\theta}_{f \text{ (products)}} - \Sigma\Delta U^{\theta}_{f \text{ (propellant components)}} \tag{8.2}$$

The amount of force exerted on the projectile by the combustion gases is dependent on the quantity and temperature of the gases, as shown in Equation 8.3,

$$F = nRT_o = \frac{RT_o}{\overline{M}} \tag{8.3}$$

where F is the force exerted on the projectile in J g^{-1}, n is the number of moles of gas produced from 1 g of propellant in mol g^{-1}, T_o is the adiabatic flame temperature of the propellant in Kelvin, $\overline{M}$ is the mean molar mass of combustion gases in g mol^{-1} and R is the universal gas constant in J mol^{-1} K^{-1}. The value F is a useful parameter for comparing the performance of gun propellants and can be determined experimentally by burning a quantity of propellant inside a 'Closed Vessel'.

Composition

The compositions of gun propellants have traditionally been fabricated from nitrocellulose-based materials. These fibrous materials have good mechanical properties and can be fabricated in granular or stick form (known as grains) to give a constant burning surface without detonation. The size of the propellant grains will depend on the size of the gun. Larger guns require larger grains which take more time to burn. The shape of the propellant grain is also very important. Grains with large surface areas will burn at a faster rate than those with low surface areas. Examples of the different geometrical shapes of propellant grains are shown in Figure 8.1.

Apart from the size and shape of the propellant grains, their composition also plays an important role. There exist three basic types of solid

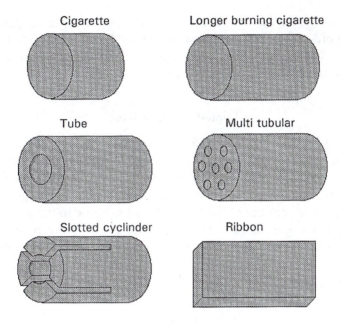

Figure 8.1 *Examples of different geometrical shapes of gun propellants*

gun propellant: these are known as single-base, double-base and triple-base. Other types of gun propellant which are less common are high energy, liquid and composite gun propellants.

Single-base Propellants

Single-base propellants are used in all kinds of guns from pistols to artillery weapons. The composition consists of 90% or more of nitrocellulose which has a nitrogen content of 12.5–13.2%. The nitrocellulose is gelled by adding a plasticizer such as carbamite or dibutyl phthalate, and then extruded and chopped into the required grain shape. The energy content (Q value) of single-base gun propellants is between 3100 and 3700 J g^{-1}.

Double-base Propellants

In order to raise the Q value of single-base propellants and to increase the pressure of the gas inside the gun barrel, nitrocellulose is mixed with nitroglycerine to form double-base propellants. Double-base propellants have a Q value of about 4500 J g^{-1} and are used in pistols and mortars. The disadvantage of double-base gun propellants is the excessive erosion of the gun barrel caused by the higher flame temperatures, and the presence of a muzzle flash which can be used to locate the gun. The muzzle flash is the result of a fuel–air explosion of the combustion products (*i.e.* hydrogen and carbon monoxide gases).

Triple-base Propellants

In order to reduce the muzzle flash in double-base propellants a third energetic material nitroguanidine is added to nitrocellulose and nitro-glycerine to form triple-base propellants. The introduction of about 50% nitroguanidine to the propellant composition results in a reduc-tion in the temperature of the flame and an increase in the gas volume. Consequently, gun barrel erosion and muzzle flash are reduced, and there is also a slight reduction in the performance of the propellant. Triple-base propellants are used in tank guns, large calibre guns and UK Naval guns.

Propellant Additives

Gun propellants contain additives which are necessary to impart certain required properties to the propellants. The additives can be classified according to their functions as shown in Table 8.1.

A given additive may be used for more than one function, for example, carbamite can be used as a stabilizer, plasticizer, coolant and surface moderant.

High Energy Propellants

In order to increase the velocity of gun propellants, nitroguanidine is replaced with RDX to form high energy propellants. High energy pro-pellants are capable of penetrating armour and are only used in tank guns where very high energies are required. The disadvantage of using high energy propellants is the extensive gun barrel erosion due to the high flame temperatures, and the vulnerability of the composition to accidental initiation which can lead to detonation.

Liquid Propellants

The advantages of using a liquid propellant compared with a solid propellant are that it is light in weight, cheap to produce, less vulnerable to accidental initiation, has a high energy output per unit volume, and has a high storage capacity.

Table 8.1 *Examples of additives used in gun propellants*

Function	Additive	Action
Stabilizer	Carbamite (diphenyl diethyl urea), methyl centralite (diphenyl dimethyl urea), chalk and diphenylamine	Increase shelf life of propellant
Plasticizer	Dibutyl phthalate, carbamite and methyl centralite	Gelation of nitrocellulose
Coolant	Dibutyl phthalate, carbamite, methyl centralite and dinitrotoluene	Reduce the flame temperature
Surface moderant	Dibutyl phthalate, carbamite, methyl centralite and dinitrotoluene	Reduce burning rate of the grain surface
Surface lubricant	Graphite	Improve flow characteristics
Flash inhibitor	Potassium sulfate, potassium nitrate, potassium aluminium fluoride and sodium cryolite	Reduce muzzle flash
Decoppering agent	Lead or tin foil, compounds containing lead or tin	Remove deposits of copper left by the driving band
Anti-wear	Titanium dioxide and talc	Reduce erosion of gun barrel

Considerable research and development has been undertaken in this area and it is estimated that a liquid gun propellant will probably come into service in the 21st century. Liquid gun propellants that are undergoing development are compositions which contain an aqueous ($\sim 63\%$) solution of the crystalline salt hydroxylammonium nitrate (HAN) and a 50:50 mixture of nitromethane and isopropyl nitrate.

Composite Propellants

Solid propellant compositions which have been previously discussed all suffer from the possibility of accidental initiation from fire, impact, electric spark, *etc.* Therefore, attention has turned to the development of insensitive munitions with particular emphasis on low vulnerability ammunition (LOVA). LOVA propellants contain RDX or HMX, an inert polymeric binder and a plasticizer. These composite propellants are less vulnerable to initiation than nitrocellulose-based propellants.

ROCKET PROPELLANTS

Performance

Rocket propellants are very similar to gun propellants in that they are designed to burn uniformly and smoothly without detonation. Gun propellants, however, burn more rapidly due to the higher operating pressures in the gun barrel. Rocket propellants are required to burn at a chamber pressure of ~ 7 MPa, compared with ~ 400 MPa for gun propellant. Rocket propellants must also burn for a longer time to provide a sustained impulse.

The specific impulse I_s is used to compare the performances of rocket propellants and is dependent on the thrust and flow rate of the gases through the nozzle as shown in Equation 8.4.

$$I_s = \frac{\text{Thrust of motor}}{\text{Mass flow rate through nozzle}} \tag{8.4}$$

The value for I_s is dependent on the velocity and pressure of the gaseous products at the nozzle exit. An expression for I_s is given in Equation 8.5,

$$I_s = \left[\frac{2F}{\gamma - 1} \left\{ 1 - \left(\frac{P_e}{P_c} \right)^{\frac{\gamma-1}{\gamma}} \right\} \right]^{1/2} \tag{8.5}$$

where F is the force constant of the propellant in J g^{-1}, γ is the ratio of specific heats for the combustion gases, P_e is the pressure at the nozzle exit and P_c is the pressure in the combustion chamber. The specific impulse is therefore dependent on the properties of the propellant and on the design of the propellant casing, *i.e.* rocket motor.

Composition

Like gun propellants, solid rocket propellants are manufactured in the form of geometrical shapes known as grains. For short range missiles the grains are larger and fewer in number than in a gun cartridge, and are designed to burn over their entire surface to give a high mass burning rate. For larger and longer distance missiles the rocket motor will only contain one or two large grains, as shown in Figure 8.2.

There are two main types of solid rocket propellant: these are composite and double-base propellants.

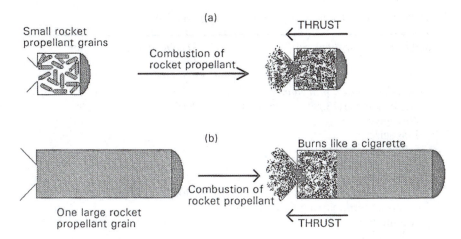

Figure 8.2 *Examples of the geometric structure of a rocket propellant for* (a) *short and* (b) *long range missiles*

Double-base Propellants

Double-base rocket propellants are homogeneous and contain nitro-cellulose plasticized with nitroglycerine. Their manufacturing process depends on the size of the propellant grain. An extruded, double-base rocket propellant results in smaller grains and is used in small rocket motors, whereas a cast, double-base rocket propellant produces larger grains and is used in large rocket motors.

Composite Propellants

Composite rocket propellants are two-phase mixtures comprising a crystalline oxidizer in a polymeric fuel/binder matrix. The oxidizer is a finely-dispersed powder of ammonium perchlorate which is suspended in a fuel. The fuel is a plasticized polymeric material which may have rubbery properties (*i.e.* hydroxy-terminated polybutadiene crosslinked with a diisocyanate) or plastic properties (*i.e.* polycaprolactone). Composite rocket propellants can be either extruded or cast depending on the type of fuel employed. For composite propellants which are plastic in nature, the technique of extrusion is employed, whereas for composite propellants which are rubbery, cast or extruded techniques are used.

Information on the performance of some solid rocket propellants is presented in Table 8.2.

Table 8.2 *Performance of some solid rocket propellants*

Solid rocket propellant	Specific impulse /N s kg^{-1}	Flame temperature/K
Double-base rocket propellant Nitrocellulose (13.25% N content) Nitroglycerine Plasticizer Other additives	2000	2500
Composite rocket propellant Ammonium perchlorate Hydroxy-terminated polybutadiene Aluminium Other additives	2400	2850
Ammonium perchlorate Carboxy-terminated polybutadiene Aluminium Other additives	2600	3500

Liquid Propellants

Liquid rocket propellants are subdivided into monopropellants and bipropellants. Monopropellants are liquids which burn in the absence of external oxygen. They have comparatively low energy and specific impulse and are used in small missiles which require low thrust. Hydrazine is currently the most widely used monopropellant; however, hydrogen peroxide, ethylene oxide, isopropyl nitrate and nitromethane have all been considered or used as monopropellants. Information on the performance of some monopropellants is presented in Table 8.3.

Bipropellants consist of two components, a fuel and an oxidizer, which are stored in separate tanks and injected into a combustion

Table 8.3 *Performance of some liquid rocket monopropellants*

Liquid monopropellant	Chemical formula	Specific impulse /N s kg^{-1}	Flame temperature/K
Hydrogen peroxide	H_2O_2	1186	900
Ethylene oxide	C_2H_4O	1167	1250
Hydrazine	N_2H_4	1863	1500
Isopropyl nitrate	$C_3H_7ONO_2$	1569	1300
Nitromethane	CH_3NO_2	2127	2400

Table 8.4 *Performance of some liquid rocket bipropellants*

Fuel	Formula	Specific impulse with different oxidizers $/N s kg^{-1}$	
		Nitrogen tetroxide	Oxygen
Hydrogen	H_2	2735	3557
Kerosene	$(CH_2)_n$	2245	2509
Methanol	CH_3OH	–	2402
Hydrazine	N_2H_4	2441	2646
UDMH*	$N_2H_2C_2H_6$	2686	2892

*UDMH – unsymmetrical dimethylhydrazine

chamber where they come into contact and ignite. The fuel component of the bipropellant includes methanol, kerosene, hydrazine, mono-methylhydrazine and unsymmetric dimethylhydrazine (UDMH). Common oxidizers are nitric acid and those based on dinitrogen tetroxide (DNTO). Some bipropellants which are gaseous at room temperature need to be stored and used at low temperatures so that they are in the liquid state. An example of this is the bipropellant mixture of hydrogen and oxygen. These types of bipropellant have very high specific impulses and are used on the most demanding missions such as satellite launch vehicles. Information on the performance of some liquid bipropellants is presented in Table 8.4.

GAS-GENERATING PROPELLANTS

Solid propellants can be used in systems where large quantities of gas are required in a very short period of time. Such systems include airbags for cars and ejector seats in aircrafts. The advantage of using propellant compositions is the speed at which the gas is generated; however, the gases are very hot, toxic and are prone to accidental initiation. Research and development is currently being undertaken in this area to produce a non-toxic gas which has a long storage life, low flame temperature and which is cheap to produce.

INTRODUCTION TO PYROTECHNICS

The name pyrotechnic is derived from the Greek words 'pyr' (fire) and 'techne' (an art), which describes the effect observed from a burning pyrotechnic composition. These effects include the production of coloured smoke, noise, and the emission of bright-coloured light. Pyrotechnic compositions are also utilized in heat-generating devices, delay and igniter compositions.

Pyrotechnics are very similar to explosive and propellant compositions. Explosives perform at the highest speed of reaction producing gaseous products, propellants are gas generators and perform at a slower speed than explosives, and pyrotechnics react at visibly observable rates with the formation of solid residues.

Pyrotechnic compositions contain a fuel and an oxidizer which is specifically formulated to produce a lot of energy. This energy is then used to produce a flame or glow (*i.e.* a matchstick), or combined with other volatile substances to produce smoke and light (*i.e.* fireworks), or to produce large quantities of gas (firework rockets and bangers).

HEAT-PRODUCING PYROTECHNICS

Heat-producing pyrotechnic compositions are used in a variety of applications, for example, as 'first fires' in pyrotechnic devices to ignite other materials, as primers, as heat generators in pyrotechnic heaters, as propellants in gas generators and rocket motors, and as incendiary devices.

Pyrotechnic compositions which are used in primers or first fires are very sensitive to initiation, whereas compositions which are used in heat-generating devices are less sensitive. The sensitivity of the compositions can be controlled by reducing the amount of oxidizer and choosing a less sensitive oxidizer.

Primers and First Fires

Pyrotechnic primer compositions are often used to ignite gun propellants. The primer emits a burst of flame when it is struck by a metal firing pin, igniting the gun propellant. These primer compositions are therefore very sensitive to initiation and are capable of generating heat and shock. In order to reduce the sensitivity during the manufacturing process, the composition is used in the paste form. Examples of pyrotechnic compositions used in primers are presented in Table 8.5.

First fire pyrotechnic compositions are used to ignite other compositions which are less sensitive to initiation. An example of a first fire composition used in fireworks is blackpowder. The blackpowder is moistened with water containing dextrin and poured on to the top of the firework composition as shown in Figure 8.3.

The fuse or blue touch-paper ignites the blackpowder which in turn ignites the main firework composition. Other types of first fire composi-

Table 8.5 *Examples of pyrotechnic compositions used as primers*

Pyrotechnic composition	Uses
Potassium chlorate Lead peroxide Antimony sulfide Trinitrotoluene	Percussion primer
Potassium perchlorate Lead thiocyanate Antimony sulfide	Stab primer

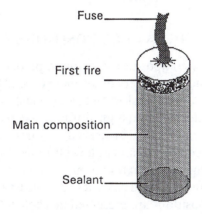

Figure 8.3 *Schematic diagram of a pyrotechnic filling for a firework*

tions include magnesium, potassium nitrate, titanium, silicon and barium peroxide. First fire compositions, like primers, are very hazardous materials to handle and are only used in small quantities.

Heat-generating Devices

Pyrotechnic compositions used in heating devices have a uniform burning rate and do not generate large amounts of energy or gas. The composition is contained in a sealed unit and produces heat without the presence of flames, sparks and gases. These heating compositions have been used in heating canned food or water, in heat cartridges for soldering irons, in activating galvanic cells which contain solid, fusible electrolytes, and in the metallurgy industry to prolong the molten state of the metal by placing packaged slabs of the heating composition on top of the molten metal. Heat-generating pyrotechnic compositions

Table 8.6 *Examples of pyrotechnic compositions used as delays*

Pyrotechnic composition	Effect
Blackpowder	Gassy
Tetranitrocarbazole and potassium nitrate	Gassy
Boron, silicon and potassium dichromate	Gasless
Tungsten, barium chromate and potassium perchlorate	Gasless
Lead chromate, barium chromate and manganese	Gasless
Chromium, barium chromate and potassium perchlorate	Gasless

contain zinc, zirconium or barium chromate, and manganese. They can be initiated by impact or friction.

DELAY COMPOSITIONS

A pyrotechnic delay composition provides a predetermined time delay between ignition and the delivery of the main effect. Delay compositions are divided into two types: gasless and gassy. Gasless delays are used in conditions of confinement or at high altitudes, where it is important that variations from normal, ambient pressure do not occur. Gassy delays are used in vented conditions and at low altitudes. Blackpowder is used as a gassy delay, whereas a mixture of metal oxides or metal chromates with an elemental fuel are used as gasless delays. Examples of gassy and gasless delay compositions are presented in Table 8.6.

The burning rate of delay compositions can be very fast (mm ms^{-1}) or quite slow (mm s^{-1}). Delay compositions which have a fast burn rate of greater than 1 mm ms^{-1} are used in projectiles and bombs that explode on impact. Delay compositions which have a slow burn rate between 1 and 6 mm s^{-1} are used in ground chemical munitions such as smoke pots, tear gas and smoke grenades. Delay compositions which have intermediate burn rates are used in effective blasting in quarries and salt mines. Here, the explosions in the boreholes are staggered to reduce vibration and improve fragmentation.

SMOKE-GENERATING COMPOSITIONS

Pyrotechnic compositions can produce a wide range of coloured smokes, such as white, black, red, green, orange, *etc.*, depending upon the formulation. These compositions produce a great deal of gas which disperses the smoke. They also burn at low temperatures so as not to degrade the organic dyes. Examples of smoke-generating pyrotechnic compositions are presented in Table 8.7.

Table 8.7 *Examples of pyrotechnic compositions used as smoke generators*

Pyrotechnic composition	Effect
Zinc dust, hexachloroethane and aluminium	White smoke
Phosphorous pentoxide and phosphoric acid	White smoke
Sulfur, potassium nitrate and pitch	Black smoke
Potassium chlorate, naphthalene and charcoal	Black smoke
Zinc dust, hexachloroethane and naphthalene	Grey smoke
Silicon tetrachloride and ammonia vapour	Grey smoke
Auramine, potassium chlorate, baking soda and sulfur	Yellow smoke
Auramine, lactose, potassium chlorate and chrysoidine	Yellow smoke
Rhodamine red, potassium chlorate, antimony sulfide	Red smoke
Rhodamine red, potassium chlorate, baking soda, sulfur	Red smoke
Auramine, indigo, potassium chlorate and lactose	Green smoke
Malachite green, potassium chlorate, antimony sulfide	Green smoke
Indigo, potassium chlorate and lactose	Blue smoke
Methylene blue, potassium chlorate, antimony sulfide	Blue smoke

In coloured smoke compositions, the volatile organic dye sublimes and then condenses in air to form small solid particles. The dyes are strong absorbers of visible light and only reflect certain discrete wavelengths of light depending upon the nature of the dye. For example, red dyes will absorb light in all regions of the visible spectrum except for the frequencies in the red region, which is reflected off the particles. Smoke-generating compositions are used in ground wind direction indicators, in flares, in screening and camouflaging, in special effects in theatres and films, and in military training aids.

LIGHT-GENERATING COMPOSITIONS

The intensity of light emitted by pyrotechnic compositions is determined by the temperature of the burning components which, in turn, is dependent on the composition. Pyrotechnic mixtures which burn between 2180 and 2250 °C contain chlorates and perchlorates as oxidizers, and an organic fuel such as shellac or rosin. In order to increase the flame temperatures to 2500-3000 °C metals powders are added such as magnesium. Light-emitting pyrotechnic compositions also contain metal compounds which produce spectral emissions at characteristic frequencies.

Coloured Light

Red light is produced by adding strontium compounds to the pyrotechnic mixture. At high temperatures the strontium compound breaks apart and reacts with the chlorine from the oxidizer [*i.e.* perchlorate (ClO_4^-) molecule] to form $SrCl^+$,* as shown in Reaction 8.1.

$$KClO_4 \rightarrow KCl + 2O_2$$

$$KCl \rightarrow K^+ + Cl^-$$

$$Cl^- + Sr^{2+} \rightarrow SrCl^{+*} \tag{8.1}$$

It is the $SrCl^+$ molecule which emits light in the red region of the electromagnetic spectrum, *i.e.* 600–690 nm. Green light is produced by adding barium compounds to the pyrotechnic mixture. Green light is emitted from the $BaCl^{+*}$ molecule at 505–535 nm. Blue light is achieved by reacting copper compounds with potassium perchlorate to form $CuCl^{+*}$ which emits light in the blue region of the visible electromagnetic spectrum, *i.e.* 420–460 nm.

White Light

White light is formed when pyrotechnic compositions burn at high temperatures. The hot, solid and liquid particles emit light in a broad range of wavelengths in the visible region of the electromagnetic spectrum resulting in white light. The intensity of the light emission is dependent on the quantity of atoms and molecules which are excited. Higher temperatures result in a substantial amount of atoms and molecules which are excited, resulting in a high intensity emission. In order to achieve these high temperatures magnesium and aluminium are used in the pyrotechnic mixture. Oxidation of these metals is a very exothermic process and a substantial amount of heat is evolved. Titanium and zirconium metals are also good elements for white light pyrotechnic compositions. Examples of pyrotechnic compositions which emit white and coloured light are presented in Table 8.8.

NOISE-GENERATING PYROTECHNICS

Bang

Pyrotechnic compositions can produce two types of noise; a loud explosive bang or a whistle. The explosive bang is achieved by placing a

---

*Other researchers believe that there is no charge on these molecules

Table 8.8 *Examples of pyrotechnic compositions which emit white and coloured light*

Pyrotechnic composition	Effect
Magnesium, barium nitrate and potassium nitrate	White light
Potassium perchlorate, barium nitrate and a binder	Green light
Potassium perchlorate, strontium oxalate and a binder	Red light
Potassium perchlorate, sodium oxalate and a binder	Yellow light
Potassium perchlorate, copper carbonate and poly(vinyl chloride)	Blue light
Magnesium, strontium nitrate and a binder	Red tracer

gas-generating pyrotechnic mixture (*i.e.* blackpowder) inside a sealed cardboard tube. The pyrotechnic mixture is ignited via a fuse and large quantities of gas are produced which eventually bursts the tube resulting in a loud bang. This principle is used in fireworks such as bangers and aerial bomb shells. A louder bang can be achieved by using a pyrotechnic flash powder. The flash powder reacts faster and at higher temperatures than blackpowder resulting in a more rapid release of high pressure gas.

Whistle

Certain pyrotechnic compositions when compressed and burnt in an open-ended tube will whistle. These compositions usually contain aromatic acids or their derivatives, potassium derivatives of benzoic acid

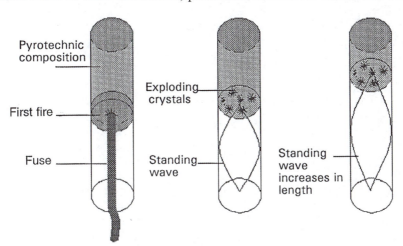

Figure 8.4 *Schematic diagram of a whistling firework*

and 2,4-dinitrophenol, picric acid and sodium salicylate. On ignition, the pyrotechnic mixture burns in an uneven manner. Small explosions occur at the burning surface which gives rise to a resonating standing wave, resulting in a whistling sound. The wavelength of the standing wave increases in length as the pyrotechnic composition reduces as shown in Figure 8.4, which results in the lowering of the frequency of the whistle.

Pyrotechnic compositions which are capable of producing a whistling noise are very sensitive and hazardous to handle.

Bibliography

J. Akhavan and M. Costello, Proceedings of the 20th International Pyrotechnic Seminar, Colorado, 1994.

J. Akhavan, Proceedings of the 22nd International Pyrotechnic Seminar, Colorado, 1996.

A. Bailey and S.G. Murray, *Explosives, Propellants and Pyrotechnics*, Brassey's (UK), Maxwell Pergamon Publishing Corporation plc, London, 1989.

F.P. Bowden and A.D. Yoffe, *Initiation and Growth of Explosion in Liquid and Solids*, Cambridge University Press, Cambridge, 1985.

K.O. Brauer, *Handbook of Pyrotechnics*, Chemical Publishing Co. Inc., New York, 1974.

J.A. Conkling, *Chemistry of Pyrotechnics, Basic Principles and Theory*, Marcel Dekker Inc., New York, 1985.

M.A. Cook, *The Science of High Explosives*, Reinhold Publishing Corporation, New York, 1958.

T.L. Davis, *The Chemistry of Powder and Explosives*, Angriff Press, Hollywood, CA, 1975.

P.E. Eaton, R.L. Gilardi and M. Zhang, *Advanced Materials* (Weinheim, Germany), V12, 2000.

H. Ellern, *Military and Civilian Pyrotechnics*, Chemical Publishing Company Inc., New York, 1968.

B.T. Fedoroff, H.A. Aaronson, O.E. Sheffield, E.F. Reese and G.D. Clift, *Encyclopedia of Explosives and Related Items, Volume 1*, Picatinny Arsenal, Dover, NJ, 1960.

B.T. Fedoroff, O.E. Sheffield, E.F. Reese and G.D. Clift, *Encyclopedia of Explosives and Related Items, Volume 2*, Picatinny Arsenal, Dover, NJ, 1962.

B.T. Fedoroff and O.E. Sheffield, *Encyclopedia of Explosives and Related Items, Volume 3*, Picatinny Arsenal, Dover, NJ, 1966.

B.T. Fedoroff and O.E. Sheffield, *Encyclopedia of Explosives and Related Items, Volume 4*, Picatinny Arsenal, Dover, NJ, 1969.

B.T. Fedoroff and O.E. Sheffield, *Encyclopedia of Explosives and Related Items, Volume 5*, Picatinny Arsenal, Dover, NJ, 1972.

B.T. Fedoroff and O.E. Sheffield, *Encyclopedia of Explosives and Related Items, Volume 6*, Picatinny Arsenal, Dover, NJ, 1974.

B.T. Fedoroff and O.E. Sheffield, *Encyclopedia of Explosives and Related Items, Volume 7*, Picatinny Arsenal, Dover, NJ, 1975.

S. Fordham, *High Explosives and Propellants*, Pergamon Press, New York, 1980.

M. Golfier, H. Graindorge, Y. Longevialle and H. Mace. Proceedings of the 29th International Annual Conference of ICT, Karlsruhe, 1998.

C.M. Hodges, J. Akhavan and M. Williams, Proceedings of the 16th International Pyrotechnic Seminar, Sweden, 1991.

Z. Jalovy, S. Zeman, M. Suceska, P. Vavra, K. Dudek and M. Rajic, *Journal of Energetic Materials*, V19, 2001.

C.H. Johansson and P.A. Persson, *Detonics of High Explosives*, Academic Press Inc., London, 1970.

S.M. Kaye, *Encyclopedia of Explosives and Related Items, Volume 8*, US Army Armament Research and Development Command, Dover, NJ, 1978.

S.M. Kaye, *Encyclopedia of Explosives and Related Items, Volume 9*, US Army Armament Research and Development Command, Dover, NJ, 1980.

S.M. Kaye, *Encyclopedia of Explosives and Related Items, Volume 10*, US Army Armament Research and Development Command, Dover, NJ, 1983.

G.F. Kinney and K.J. Graham, *Explosive Shocks in Air*, Springer Verlag New York Inc., New York, 1985.

K.K. Kuo and M. Summerfield, *Fundamentals of Solid-Propellant Combustion, Volume 90*, Progress in Astronautics and Aeronautics, American Institute of Aeronautics and Astronautics Inc., New York, 1984.

C. Leach, S.B. Langston and J. Akhavan, Proceedings of the 25th International Annual Conference of ICT, Karlsruhe, 1994.

R. Meyer, *Explosives*, VCH Publishers, New York, 1987.

S.J.P. Palmer, J.E. Field and J.M. Huntley, *Proc. Roy. Soc.*, London, 1993, vol. 440, pp. 399–419.

T. Shimizu, *Fireworks, The Art, Science and Technique*, Maruzen Co Ltd, Tokyo, Japan, 1981.

G. Singh and P.S.Felix, *Journal of Hazardous Materials*, A90, 2002.

G. Singh, I.P.S. Kapoor, S.K.Tiwari and P.S.Felix, *Journal of Hazardous Materials* B81, 2001.

T. Urbanski, *Chemistry and Technology of Explosives, Volume 1*, Pergamon Press Ltd, London, 1988.

T. Urbanski, *Chemistry and Technology of Explosives, Volume 3*, Pergamon Press Ltd, London, 1990.

T. Urbanski, *Chemistry and Technology of Explosives, Volume 4*, Pergamon Press Ltd, London, 1984.

Subject Index